2001 | General

[BLANK PAGE]

Standard Grade | General | Credit

Geography

General Level 2001

Credit Level 2001

General Level 2002

Credit Level 2002

General Level 2003

Credit Level 2003

General Level 2004

Credit Level 2004

General Level 2005

Credit Level 2005

Leckie×Leckie

© **Scottish Qualifications Authority**

All rights reserved. Copying prohibited. No part of this publication may be reproduced, stored in a retrieval system, or transmitted in any form or by any means, electronic, mechanical, photocopying, recording or otherwise.

First exam published in 2001.
Published by Leckie & Leckie, 8 Whitehill Terrace, St. Andrews, Scotland KY16 8RN tel: 01334 475656 fax: 01334 477392
enquiries@leckieandleckie.co.uk www.leckieandleckie.co.uk
ISBN 1-84372-301-8
A CIP Catalogue record for this book is available from the British Library.
Printed in Scotland by Scotprint.
Leckie & Leckie is a division of Granada Learning Limited, part of ITV plc.

Acknowledgements

Leckie & Leckie is grateful to the copyright holders, as credited at the back of the book, for permission to use their material.
Every effort has been made to trace the copyright holders and to obtain their permission to use their copyright material.
Leckie & Leckie will gladly receive information enabling them to rectify any error or omission in subsequent editions.

Official SQA Past Papers: General Geography 2001

FOR OFFICIAL USE

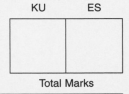

1260/403

NATIONAL QUALIFICATIONS 2001

WEDNESDAY, 23 MAY
G/C 9.00 AM – 10.25 AM
F/G 10.25 AM – 11.50 AM

**GEOGRAPHY
STANDARD GRADE**
General Level

KU	ES

Total Marks

Fill in these boxes and read what is printed below.

Full name of centre

Town

Forename(s)

Surname

Date of birth
Day Month Year

Scottish candidate number

Number of seat

1 Read the whole of each question carefully before you answer it.

2 Write in the spaces provided.

3 Where boxes like this ☐ are provided, put a tick ✓ in the box beside the answer you think is correct.

4 Try all the questions.

5 Do not give up the first time you get stuck: you may be able to answer later questions.

6 Extra paper may be obtained from the invigilator, if required.

7 Before leaving the examination room you must give this book to the invigilator. If you do not, you may lose all the marks for this paper.

Official SQA Past Papers: General Geography 2001

Extract No 1212/64

1:50 000 Scale
Landranger Series

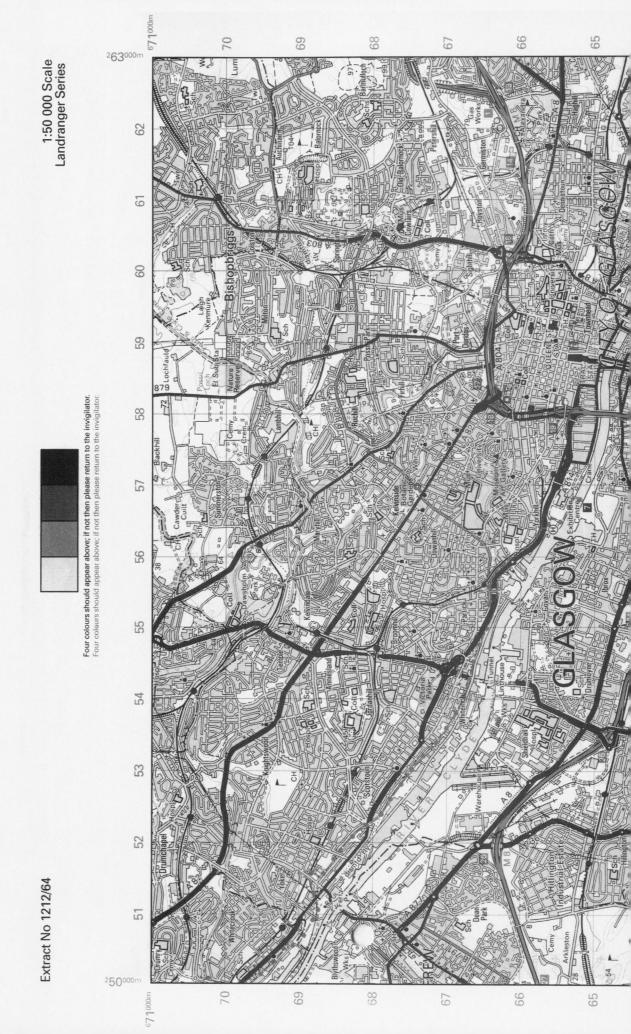

Official SQA Past Papers: General Geography 2001

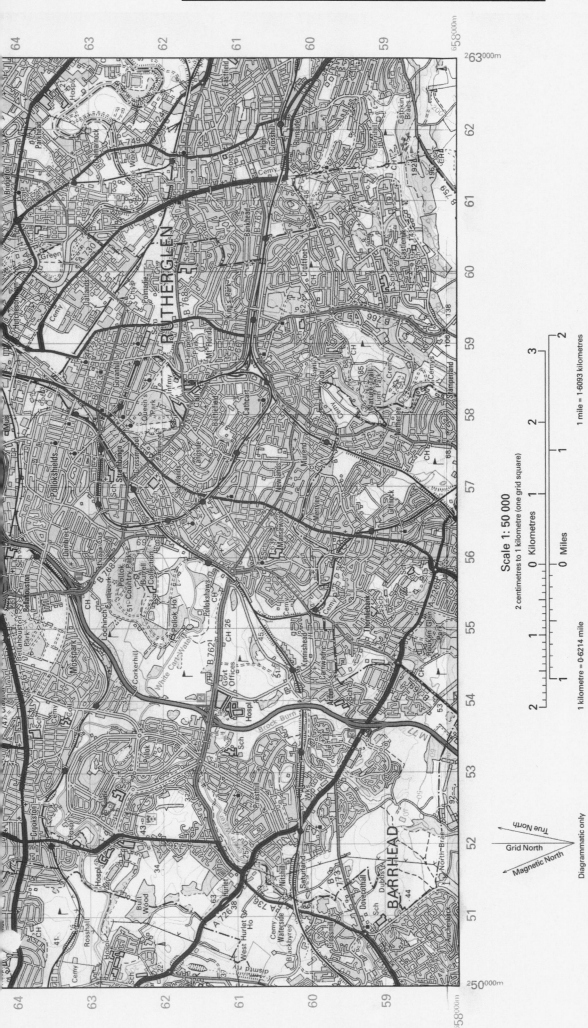

1. **Reference Diagram Q1A**

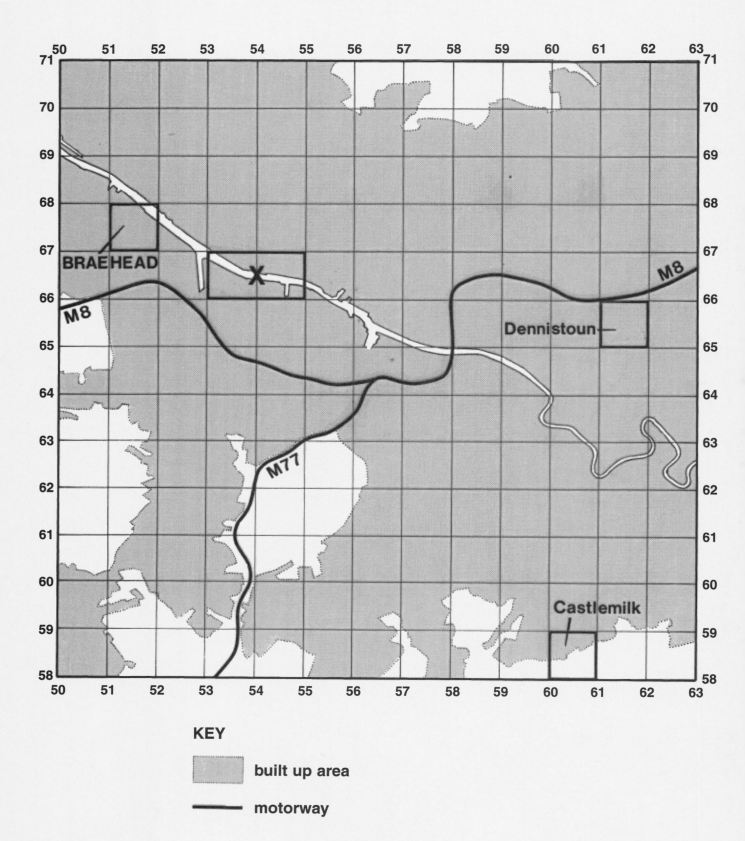

KEY

built up area

motorway

1. (continued)

Look at the Ordnance Survey Map Extract (No 1212/64) of the central Glasgow area and Reference Diagram Q1A on *Page two*.

(a) Dennistoun (6165) and Castlemilk (6058) are both residential areas.

Using **map evidence**, describe the **differences** between these two areas.

Dennistoun has better transport links as it's close to a main motorway where as castle milk is very isolated. on the edge of town. Dennistoun is very close to the river

(b) **Reference Diagram Q1B: Braehead Shopping Centre**

Reference Diagram Q1B shows the recently built shopping centre at Braehead (5167).

Using **map evidence**, give reasons why it was built at this location.

It's right by the river which makes it very appealing to tourists/visitors

[Turn over

1. (continued)

(c) Look at Reference Diagram Q1A and the map extract.

Give **map evidence** to show that area X (5366 and 5466), shown on Reference Diagram Q1A, is an industrial zone.

_____

_____

_____

_____

_____

4

(d) Reference Diagram Q1A shows the extension to the M77, opened in 1997.

Using **map evidence**, give the advantages **and** disadvantages of this new motorway.

Advantages _____

_____

_____

_____

Disadvantages _____

_____

_____

_____

4

1. **(continued)**

   (e) **Reference Text Q1C:
   Selected Human Activities on River Clyde**

   - Industry
   - Communications
   - Housing
   - Recreation

   Look at Reference Text Q1C and the map extract.

   In what ways has the River Clyde both encouraged **and** restricted human activities in Glasgow?

   _____

   _____

   _____

   _____

   4

   [Turn over

2. **Reference Diagram Q2: Ox-Bow Lake**

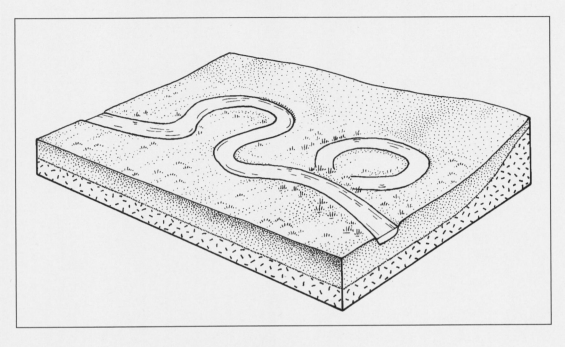

Explain how an ox-bow lake is formed.

You may use diagrams to illustrate your answer.

3

3. Reference Diagram Q3A: Selected Climate Regions

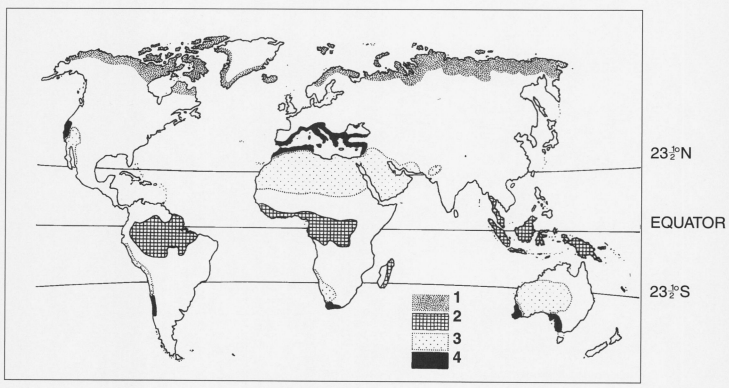

(a) Look at Reference Diagram Q3A.

Complete the table below by naming the climate regions **1** to **4** shown on the map.

Choose from:

Mediterranean, Tundra, Hot Desert, Equatorial Rain Forest.

Number	Climate Region
1	
2	
3	
4	

3. (continued)

Reference Diagram Q3B: Climate Graph

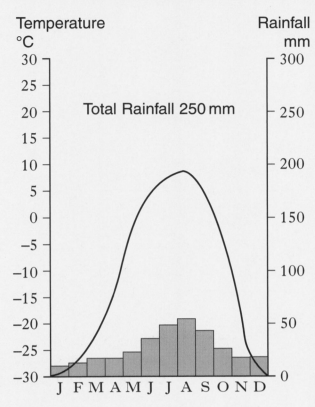

(b) Look at Reference Diagram Q3B.

The climate graph shows one of the four climates shown on Reference Diagram Q3A.

(i) Identify the climate shown by the graph.

Climate _____

(ii) Give reasons for your choice.

Reasons _____

4. **Reference Diagram Q4A: Antarctica**

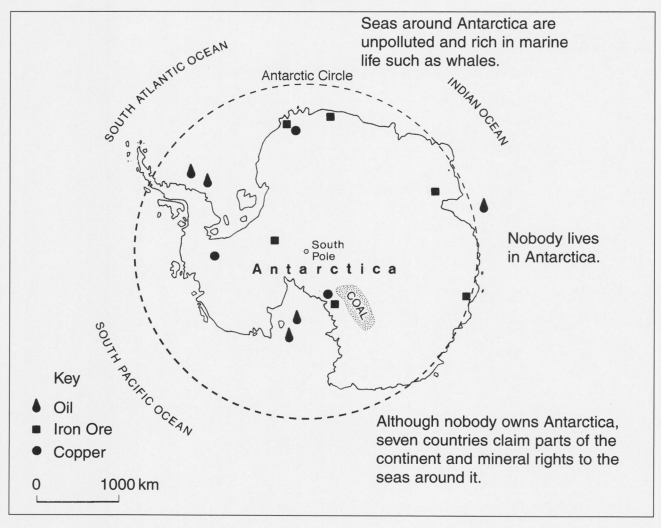

Reference Diagram Q4B: Different Views about the Antarctic Region

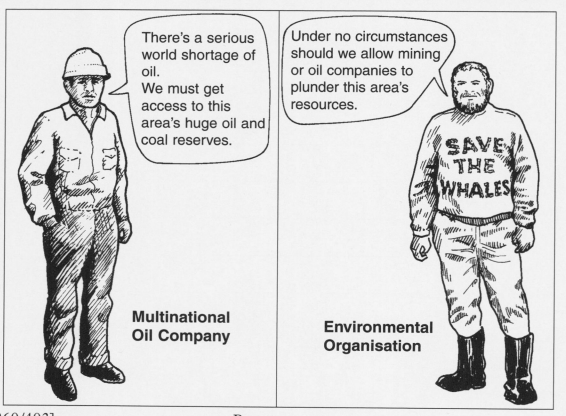

4. (continued)

Look at Reference Diagrams Q4A and Q4B.

Do you think Antarctica's mineral resources should be developed?

Give reasons for your answer.

Tick (✓) your choice.   YES ☐   NO ☐

Reasons _____

_____

_____

_____

_____

_____

_____

4

[**Turn over**

5. **Reference Diagram Q5A: Farming Landscape in 1950**

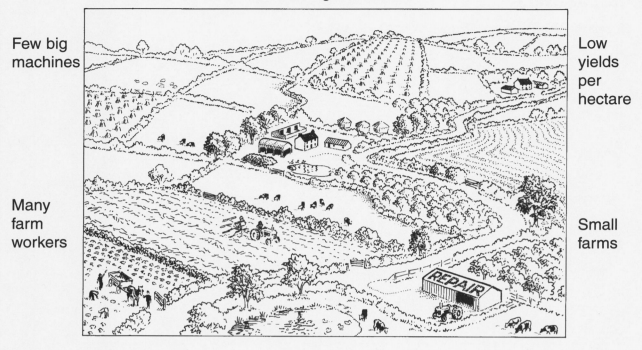

Reference Diagram Q5B: Farming Landscape in 2000

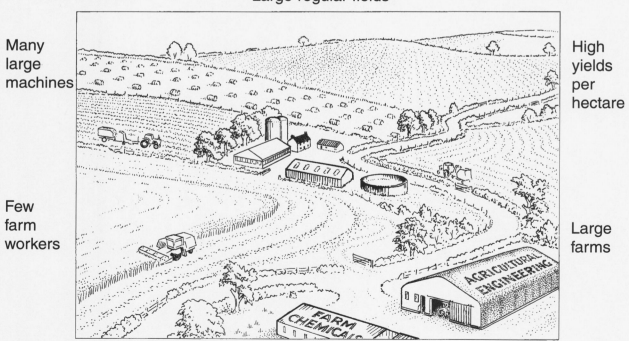

5. (continued)

(a) Look at Reference Diagrams Q5A and Q5B.

What are the advantages **and** disadvantages of the changes which have taken place in farming since 1950?

Advantages _____

_____

_____

Disadvantages _____

_____

_____

4

[Turn over

5. (continued)

Reference Table Q5C: Farm Information collected by Student

Field Number	Field Size (hectares)	Slope Steepness (degrees)	Land Use
1	5	2	barley
2	7	12	permanent grass
3	8	4	potatoes
4	12	6	barley
5	13	20	rough grazing

Reference Diagram Q5D: Map of Fields on Farm

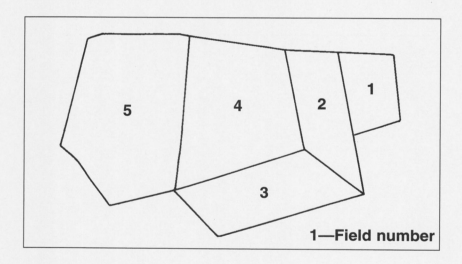

1—Field number

(b) Look at Reference Table Q5C and Reference Diagram Q5D.

Choose two **different** techniques to process the farm data that the student has collected.

Technique 1 _____

Technique 2 _____

Justify your choices. _____

4

6. Reference Diagram Q6: Ben Lawers Area

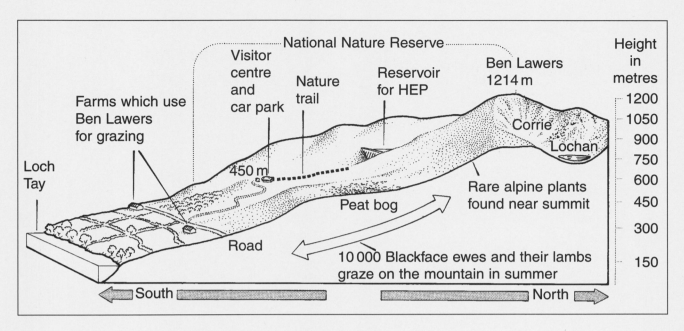

Look at Reference Diagram Q6.

(a) Many people visit Ben Lawers. What problems might this create for the area?

_____
_____
_____
_____

4

(b) A group of secondary pupils is to investigate the relationship between land use and height on Ben Lawers.

Describe **two** techniques which they could use to collect appropriate data.

Technique 1 _____

Technique 2 _____

Justify your choices. _____
_____
_____

4

7. Reference Diagram Q7: Newspaper Headlines of the 1980s

| City unemployment rate tops 40% | Massive redundancies in shipbuilding and textiles |

Look at Reference Diagram Q7.

Since the 1980s, the decline of traditional industry such as shipbuilding, textiles and coal mining has caused many problems.

What is being done to overcome these problems?

_____

_____

_____

_____

_____

3

8. Reference Diagram Q8: Projected Changes in World Population

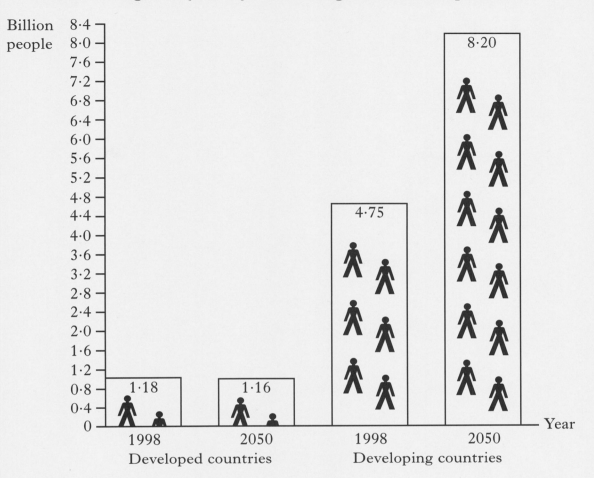

(a) Look at Reference Diagram Q8.

Describe in detail the changes in population predicted for the developed and developing countries.

(b) Describe the problems which **developing** countries are likely to have as a result of the changes in their population.

9. (a) **Reference Table Q9A: Japan—Main Export Partners**

Country	Percentage of Trade
USA	80
Newly industrialised countries, eg South Korea	10
European Union	5
Australia	5

Use the information in Reference Table Q9A to complete the pie chart below (Reference Diagram Q9B).

Reference Diagram Q9B

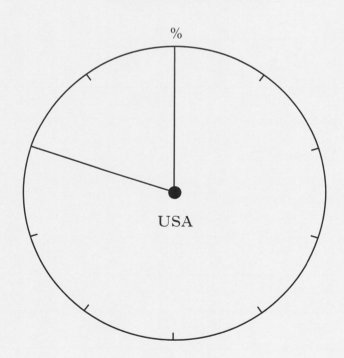

9. **(continued)**

(b) **Reference Table Q9C: Japan—Key Statistics**

Selected Exports	World Rank
Car Manufacturing	1
Computer Chips	1
Telecommunications	1
Shipbuilding	3
Iron and Steel	3

Population (millions) 2000	130
Population (world rank)	7
GNP (world rank)	2

Look at the Reference Table Q9C.

Give reasons why Japan is one of the world's economic superpowers.

_____
_____
_____
_____
_____

4

[Turn over for Question 10 on *Page twenty*]

10. **Reference Text Q10A: Problems of a Village in Mali, West Africa**

- very few children go to school
- mothers have to walk miles for water
- most people cannot read or write
- no rain, so the crops have died
- many babies ill or dying of hunger

Reference Text Q10B: Selected Types of Aid

| Send emergency food and medicine | Set up a local school with trained teachers | Provide irrigation scheme |

A   B   C

Look at Reference Texts Q10A and Q10B above.

What type of aid do you think would be most suited to this village?

Tick (✓) your choice.

A ☐   B ☐   C ☐

Give reasons for your answer.

_____
_____
_____
_____

4

[END OF QUESTION PAPER]

2001 | Credit

1260/405

NATIONAL QUALIFICATIONS 2001

WEDNESDAY, 23 MAY 10.45 AM – 12.45 PM

GEOGRAPHY
STANDARD GRADE
Credit Level

All questions should be attempted.

Candidates should read the questions carefully. Answers should be clearly expressed and relevant.

Credit will always be given for appropriate sketch-maps and diagrams.

Write legibly and neatly, and leave a space of about one cm between the lines.

Marks may be deducted for bad spelling and bad punctuation, and for writing that is difficult to read.

All maps and diagrams in this paper have been printed in black only: no other colours have been used.

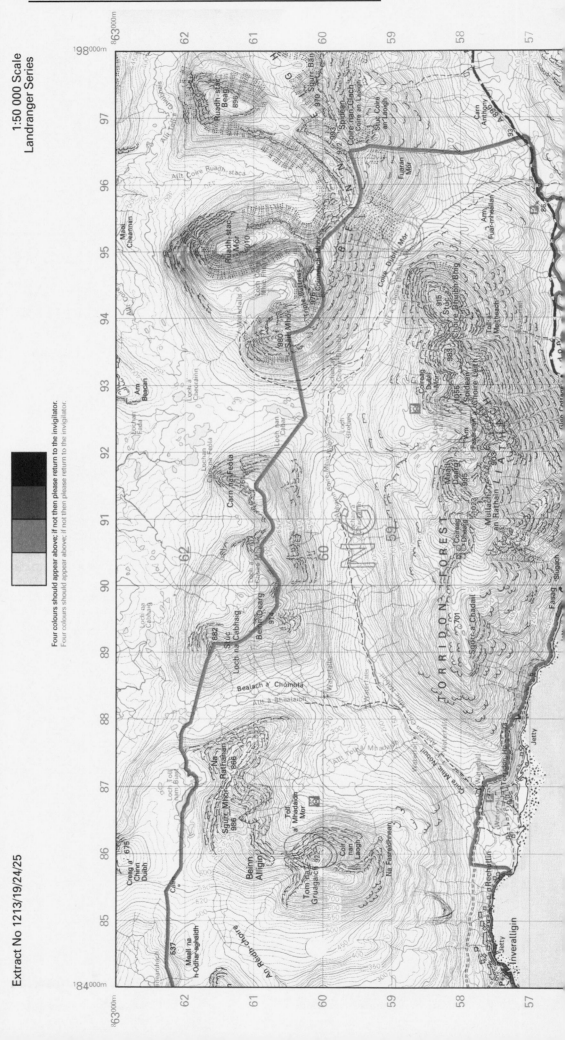

Official SQA Past Papers: Credit Geography 2001

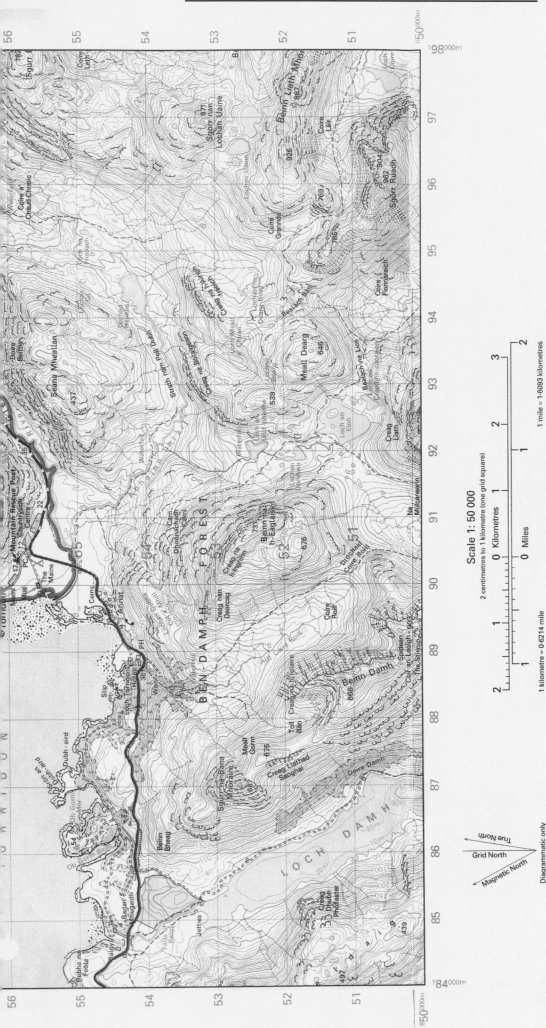

1. **Reference Diagram Q1A**

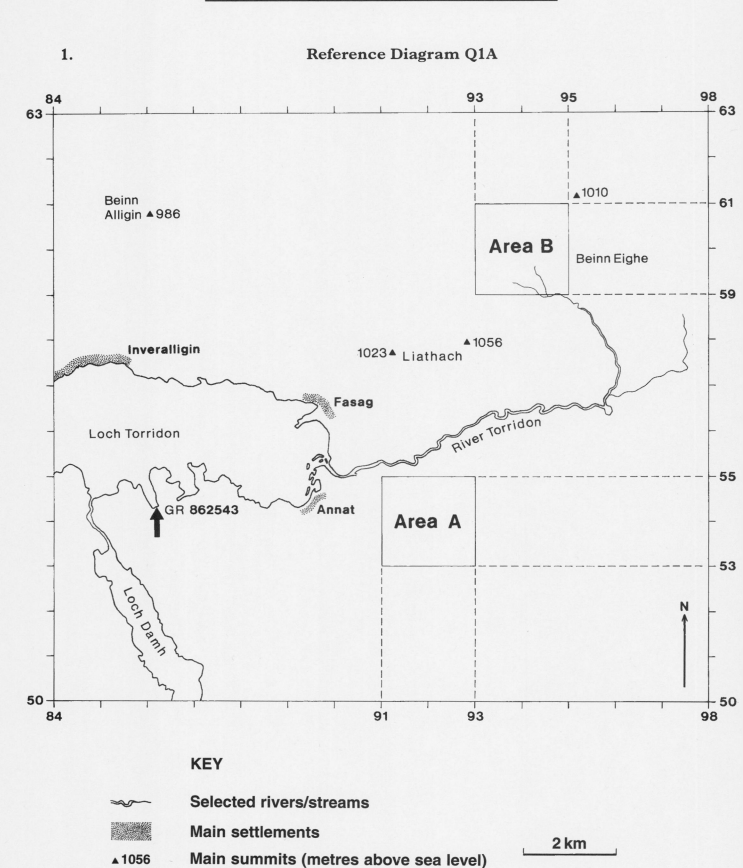

1. (continued)

This question refers to the OS Map Extract (No 1213/19/24/25) of Torridon and Reference Diagram Q1A on *Page two*.

(a) The area covered by the map extract is one of Scotland's most spectacular mountain landscapes.

   (i) Match each of the features named below with the correct grid reference.

   Features: pyramidal peak, hanging valley, arete, corrie

   Choose from:

   Grid references: 952588, 923580, 852585, 860601, 925576

   (ii) **Explain** how **one** of the features listed in (*a*)(i) was formed. You may use diagrams to illustrate your answer.

(b) Look at Reference Diagram Q1A.

   A commercial forestry company surveyed the map area's potential for forestry. It considered Area A to be more suitable than Area B.

   Using map evidence, suggest **three** reasons for this.

Reference Diagram Q1B: A Hydro-electric Power Scheme

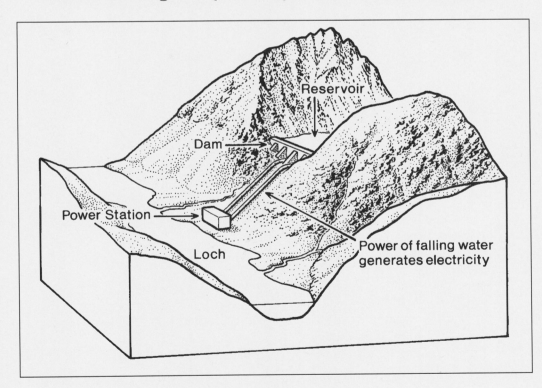

(c) Look at Reference Diagram Q1B.

   There is a plan to construct a similar dam and power station in grid square 8853.

   Using map evidence, describe the advantages **and** disadvantages of building such a scheme at this site.

1. (continued)

   Reference Text Q1C: Report of Economic Survey

   > "Torridon is an ideal place for a large scale tourist development. The benefits to the area would outweigh the disadvantages."

   Reference Diagram Q1D: The Location of Torridon

   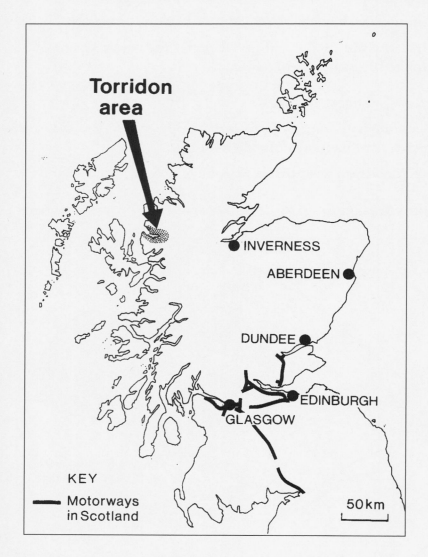

   (d) Look at Reference Text Q1C, Reference Diagram Q1D and the map extract. Do you agree with the opinion being expressed by the authors of the report? Making reference to map evidence, give reasons for your answer.

1. (continued)

Reference Diagram Q1E: Field Sketch looking North to Ben Alligin from GR 862543

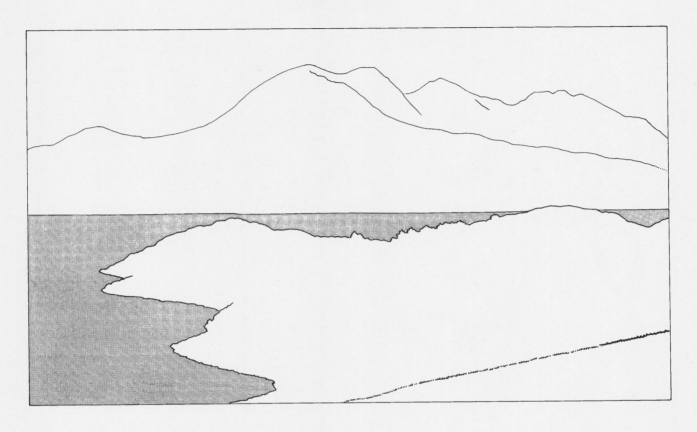

(e) A student has gathered information for a physical landscape investigation.

The information includes the outline field sketch shown above in Reference Diagram Q1E, a photograph of the same view, a geology map and the OS map extract.

What techniques should the student now use to **process** this information?

Justify your choices.

4

[Turn over

2. Reference Diagram Q2A: Loch Lomond and the Trossachs

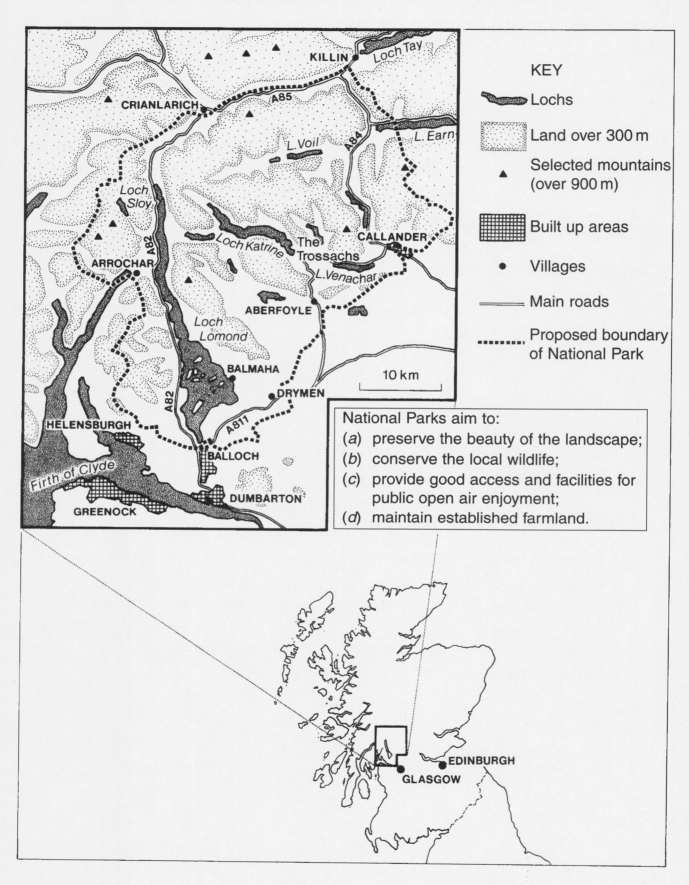

2. (continued)

Reference Diagram Q2B: Land Users in the Loch Lomond and Trossachs Area

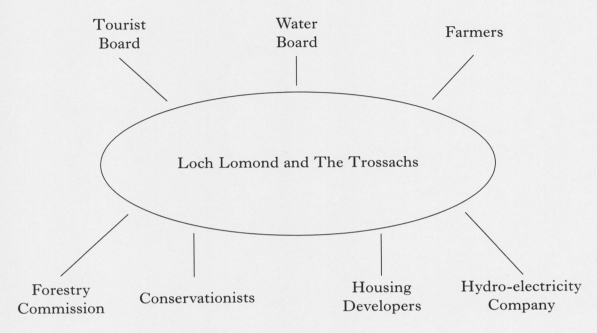

Study Reference Diagrams Q2A and Q2B.

Loch Lomond and The Trossachs was chosen as Scotland's first National Park.

Do you think all land users in Reference Diagram Q2B will welcome the establishment of this National Park?

Explain your answer.

6

[Turn over

3. **Reference Diagram Q3: Synoptic Chart for British Isles at 0700 hours on 31 August**

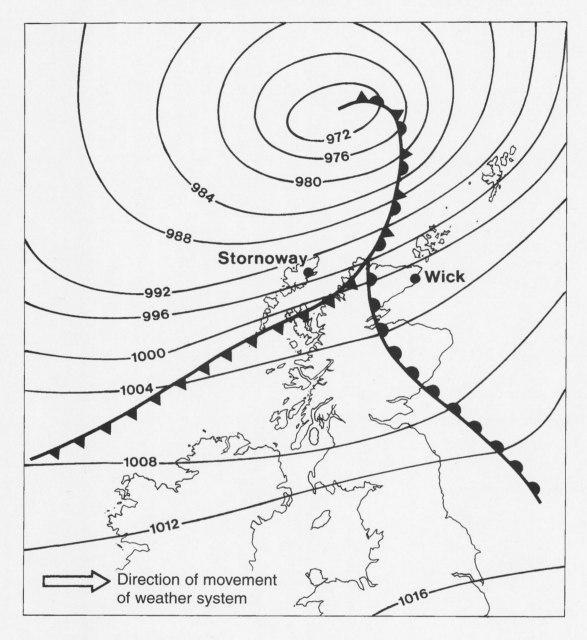

Look at Reference Diagram Q3.

A yacht race from Wick to Stornoway was due to start from Wick harbour at 8.00 am on 31 August.

At 7.00 am the Meteorological (Met) Office advised the race organisers to cancel the race.

With reference to the synoptic chart, **explain** why this advice was given.

4. Reference Diagram Q4A: Urban Transect along Queen's Road, Aberdeen

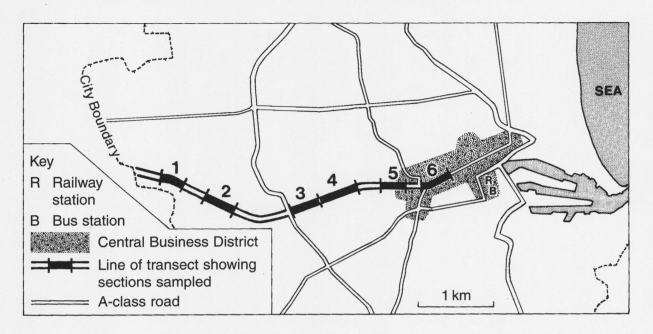

Reference Diagram Q4B: Data collected along Transect

Look at Reference Diagrams Q4A and Q4B.

(a) What techniques could have been used to gather the information in Reference Diagram Q4B?

Give reasons for your choices.

(b) **Explain** the changes that occur along the transect from the edge of the city to the centre.

5. **Reference Diagram Q5A: Examples of High-Technology Industry**

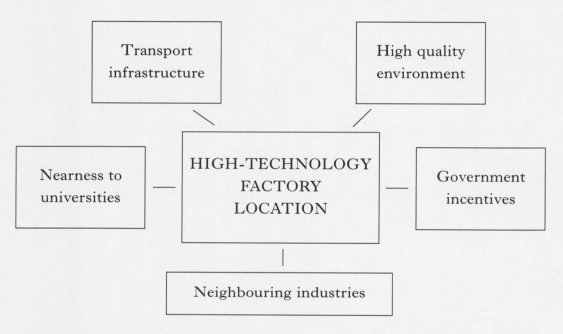

Reference Diagram Q5B: Selected Industrial Location Factors

Study Reference Diagrams Q5A and Q5B.

Explain in what ways the factors listed above are important in the location of high-technology industries.

6. **Reference Diagram Q6: Births per Woman and Infant Mortality in Selected Countries**

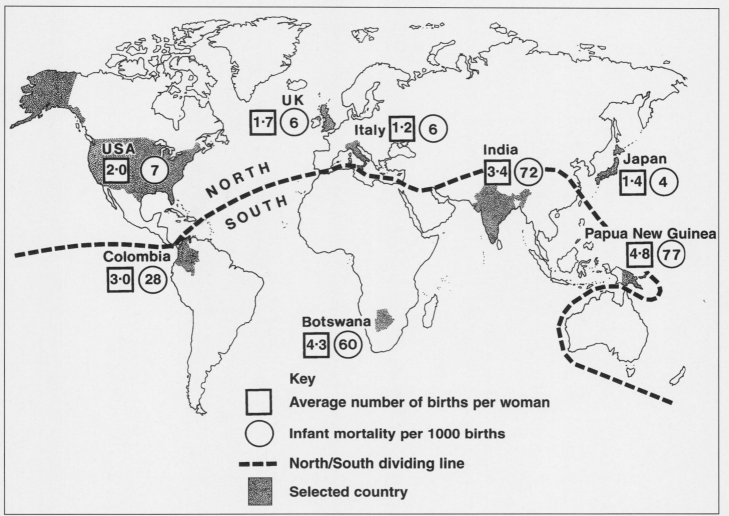

(a) Look at Reference Diagram Q6.

Describe the patterns of births per woman and infant mortality, as shown in Reference Diagram Q6. **4**

(b) Explain the differences in the patterns between North and South. **3**

(c) What measures have **developing** nations taken to reduce population growth **and** infant mortality rates? **4**

[Turn over

7. Reference Diagram Q7A:
   Japan's Population Pyramid 1950

Reference Diagram Q7B:
Japan's Population Pyramid 2050
(projected figures)

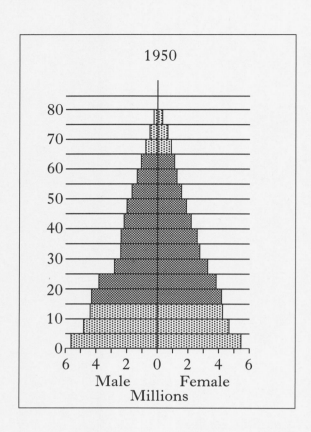
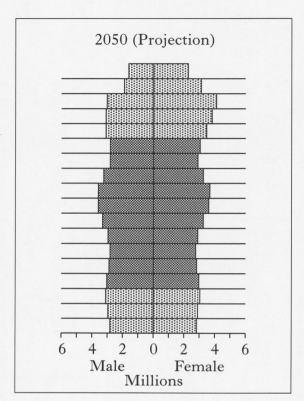

(a) Look at Reference Diagrams Q7A and Q7B.

Describe in detail the changes in Japan's population structure between 1950 and 2050.

(b) Do you agree that the changes in population structure will cause problems for the Japanese government by 2050?

State yes **or** no, and give reasons for your answer.

8. **Reference Table Q8A: Kenya—Exports and Imports**

Exports	%	Imports	%
Foodstuffs	59	Manufactured goods	40
Minerals and fuels	21	Minerals and fuels	23
Machinery and transport	9	Machinery and transport	13
Chemicals	4	Chemicals	11
Manufactured goods	3	Foodstuffs	9
Others	4	Others	4
Total ($ million)	$1028	Total ($ million)	$2136

Reference Table Q8B: Kenya—Direction of Trade

Trading Partners	% of Exports	% of Imports
European Union	43	43
(including UK)	(16)	(16)
African countries	18	2
United Arab Emirates	—	12
Japan	3	12
USA	4	4
Others	32	27

(a) Look at Reference Tables Q8A and Q8B.

Describe the pattern of Kenya's trade. **4**

(b) Describe methods you could use to process the information shown in Reference Tables Q8A and Q8B.

Justify your choices. **6**

[END OF QUESTION PAPER]

2002 | General

Official SQA Past Papers: General Geography 2002

FOR OFFICIAL USE

G

1260/403

KU | ES

Total Marks

NATIONAL QUALIFICATIONS 2002

MONDAY, 13 MAY 10.25 AM–11.50 AM

GEOGRAPHY STANDARD GRADE
General Level

Fill in these boxes and read what is printed below.

Full name of centre

Town

Forename(s)

Surname

Date of birth
Day Month Year

Scottish candidate number

Number of seat

1 Read the whole of each question carefully before you answer it.

2 Write in the spaces provided.

3 Where boxes like this ☐ are provided, put a tick ✓ in the box beside the answer you think is correct.

4 Try all the questions.

5 Do not give up the first time you get stuck: you may be able to answer later questions.

6 Extra paper may be obtained from the invigilator, if required.

7 Before leaving the examination room you must give this book to the invigilator. If you do not, you may lose all the marks for this paper.

1. Reference Diagram Q1A

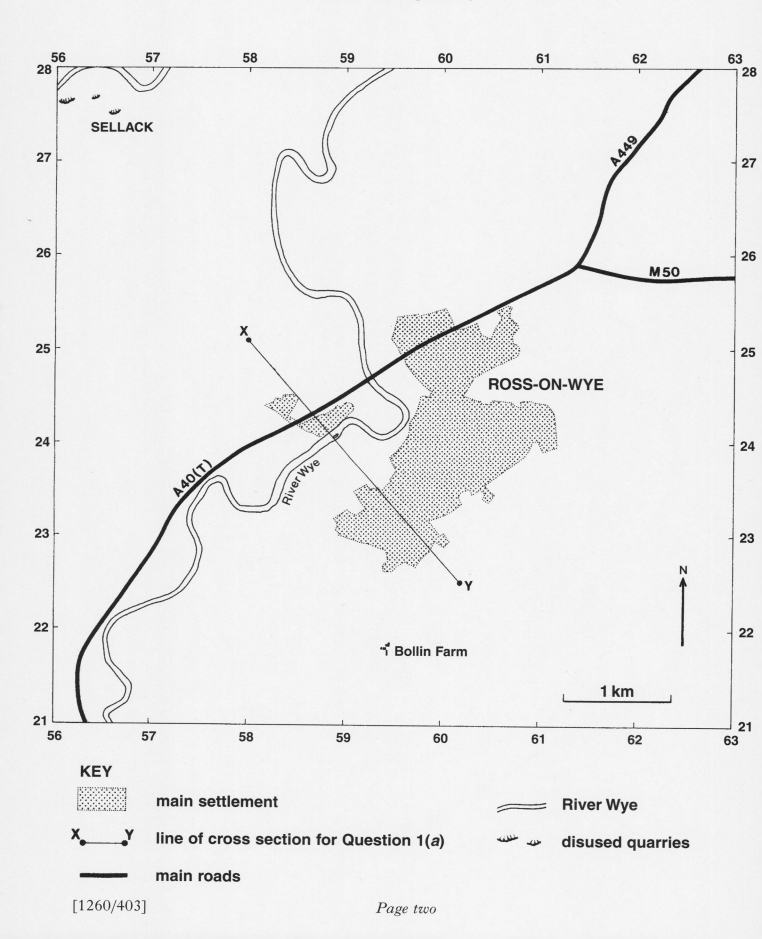

1. (continued)

Look at the Ordnance Survey Map Extract (No 1266/OLM14) of the Ross-on-Wye area and Reference Diagram Q1A on *Page two*.

Reference Diagram Q1B: Cross-section from Claytons (580251) to the Fort (602225)

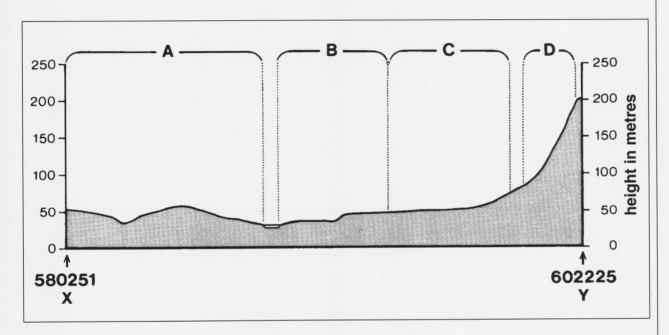

(a) Look at Reference Diagrams Q1A, Q1B and the OS Map.

Four areas of land use have been identified on the cross-section with letters A, B, C and D.

For each of the land uses listed below, write the correct letter from the cross-section in the space provided.

Descriptions	Cross-section Letter
Outskirts of town	
Woodland	
Villages and farm land	
Pasture beside the river	

3

[Turn over

1. (continued)

(b) Describe the **physical** features of the River Wye **and** its valley from where the river enters the map at 594280 to Wilton Bridge (590243).

_____

_____

_____

_____

_____

_____

3

(c) Describe **two** techniques which could be used to gather information about the **physical** characteristics of the River Wye.

_____

_____

_____

Why are these techniques suitable?

_____

_____

_____

4

1. **(continued)**

   (d) Which type of farm is Bollin Farm 594218 likely to be?

   Tick (✓) your choice.

   Livestock ☐ Arable ☐ Mixed ☐

   Give reasons to support your answer.

   _____
   _____
   _____
   _____
   _____

   (e) It is proposed to re-open and extend the quarries near Sellack in grid square 5627.

   Do you agree or disagree with the proposal?

   Give reasons for your answer.

   _____
   _____
   _____
   _____
   _____

[Turn over

1. (continued)

(f) Using map evidence, describe the advantages **and** disadvantages of the site of the settlement of Ross-on-Wye.

Advantages _____

_____

_____

_____

Disadvantages _____

_____

_____

_____

4

(g) What is the **main** function of Ross-on-Wye?

Tick (✓) your choice.

Holiday resort ☐   Industrial town ☐   Market town ☐

Give map evidence to support your choice.

_____

_____

_____

_____

_____

_____

4

[Turn over for Question 2 on *Page eight*

2. Reference Diagram Q2: Area of Glacial Deposition

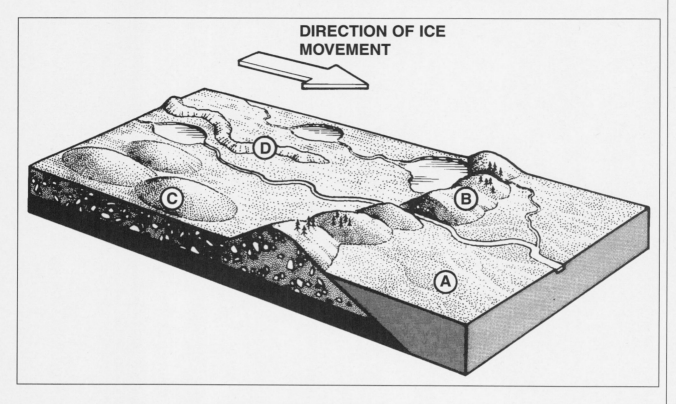

Look at Reference Diagram Q2.

(a) Match each of the features of glacial deposition in the table to the correct letter (A, B, C, D) on the Reference Diagram.

Feature	Letter
Drumlin	
Outwash Plain	
Esker	
Terminal Moraine	

3

DO NOT WRITE IN THIS MARGIN

KU	ES

Marks

2. (continued)

(*b*) **Explain** how **one** of these features was formed. You may use a diagram to illustrate your answer.

Feature ____________________

Explanation ____________________

____

____

____

____

____

____

____

____

____

____________________ **3**

[Turn over

3.

Reference Diagram Q3A: Selected Climate Regions

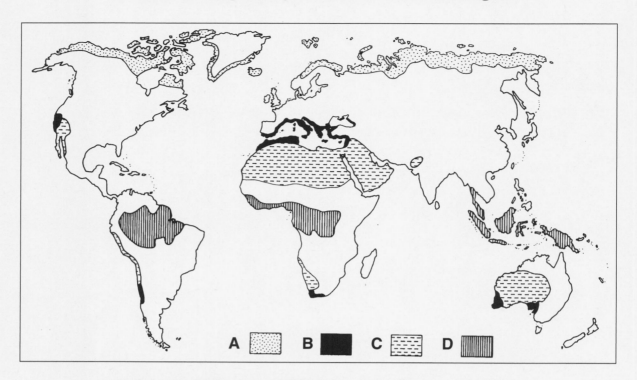

Reference Diagram Q3B: Selected Climate Graphs

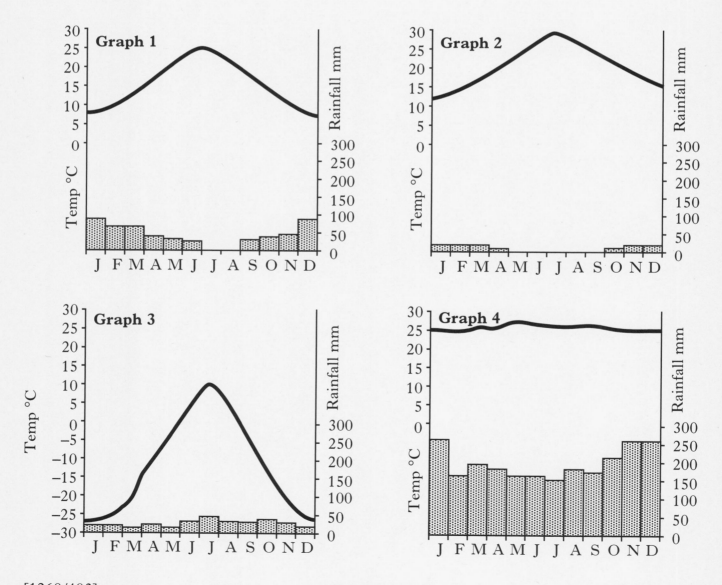

3. (continued)

Look at Reference Diagrams Q3A and Q3B.

(a) Complete the table below by adding the appropriate letter or number.

Climate Type	Map Area	Graph
Hot Desert	C	
Equatorial Rainforest		4
Mediterranean	B	
Tundra		3

(b) Describe **in detail** the main features of the climate shown in **Graph 3**.

_____
_____
_____
_____
_____

[Turn over

4. Reference Diagram Q4: **Desertification**

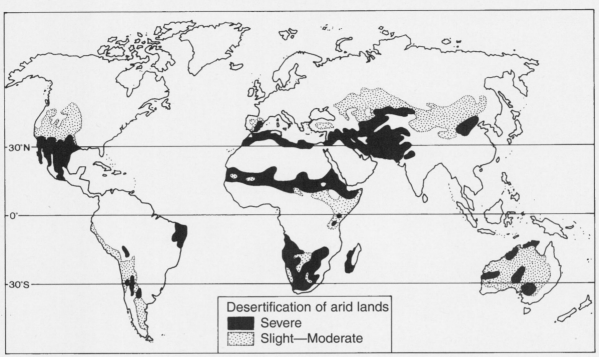

What are the main causes of desertification?

5. Reference Diagram Q5A: Relief in Scotland

Reference Diagram Q5B: January Temperatures in Scotland

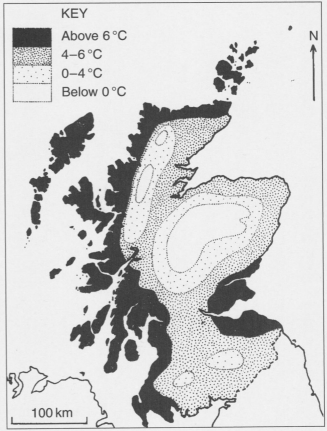

Look at Reference Diagrams Q5A and Q5B.

Describe in detail the relationship between relief and January temperatures in Scotland.

3

6. Reference Diagram Q6: Two Housing Areas in Liverpool

Area A 19th Century Housing   **Area B 20th Century Housing**

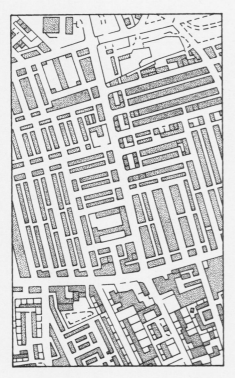 

Scale: 100 m

Look at the areas A and B above.

(a) Describe the differences in housing density and street patterns between the two areas.

DO NOT WRITE IN THIS MARGIN

KU ES

Marks

6. (continued)

(*b*) Describe **two** techniques you could use to gather information in areas A and B to show differences in quality of environment.

Give reasons for your choice.

4

[Turn over

7. **Reference Diagram Q7: Doxford International Business and Technology Park**

7. (continued)

(a) A business and technology park is a planned industrial area for offices and hi-tech industries.

Study Reference Diagram Q7.

Describe the advantages of the Doxford International Business and Technology Park **for the companies** located there.

4

(b) Describe the advantages **for older industrial areas** such as Sunderland of having many modern companies located nearby.

4

[Turn over

8. **Reference Table Q8: Urban Population as a percentage of Total Population**

	1950	1970	1990	2000
India	17	20	28	34

Look at Reference Table Q8.

(a) Give reasons for the changes in the percentage of urban population of Developing countries such as India.

4

(b) Give **one** processing technique which could be used to show the information in Reference Table Q8. Give reasons for your choice.

3

9. Reference Diagram Q9: Government considering entry to a Trade Alliance

Look at Reference Diagram Q9.

What are the advantages **and** disadvantages for countries which are members of a Trading Alliance?

4

[Turn over for Question 10 on *Page twenty*

10. Reference Diagram Q10: Examples of Intermediate Technology

Village biogas plant takes all waste and provides gas for fuel and fertiliser for crops.

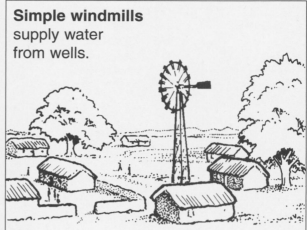
Simple windmills supply water from wells.

Look at Reference Diagram Q10.

Explain why intermediate technology, such as that shown in Reference Diagram Q10, is suitable for many rural communities in Developing countries.

4

[END OF QUESTION PAPER]

2002 | Credit

1260/405

NATIONAL
QUALIFICATIONS
2002

MONDAY, 13 MAY
1.00 PM – 3.00 PM

GEOGRAPHY
STANDARD GRADE
Credit Level

All questions should be attempted.

Candidates should read the questions carefully. Answers should be clearly expressed and relevant.

Credit will always be given for appropriate sketch-maps and diagrams.

Write legibly and neatly, and leave a space of about one cm between the lines.

Marks may be deducted for bad spelling and bad punctuation, and for writing that is difficult to read.

All maps and diagrams in this paper have been printed in black only: no other colours have been used.

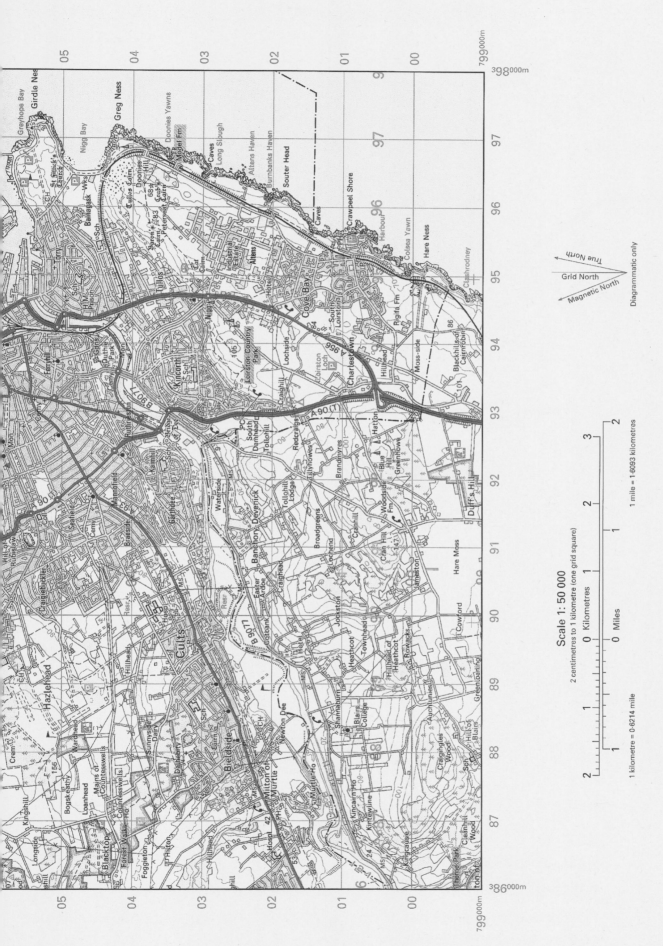

1. **Reference Diagram Q1A**

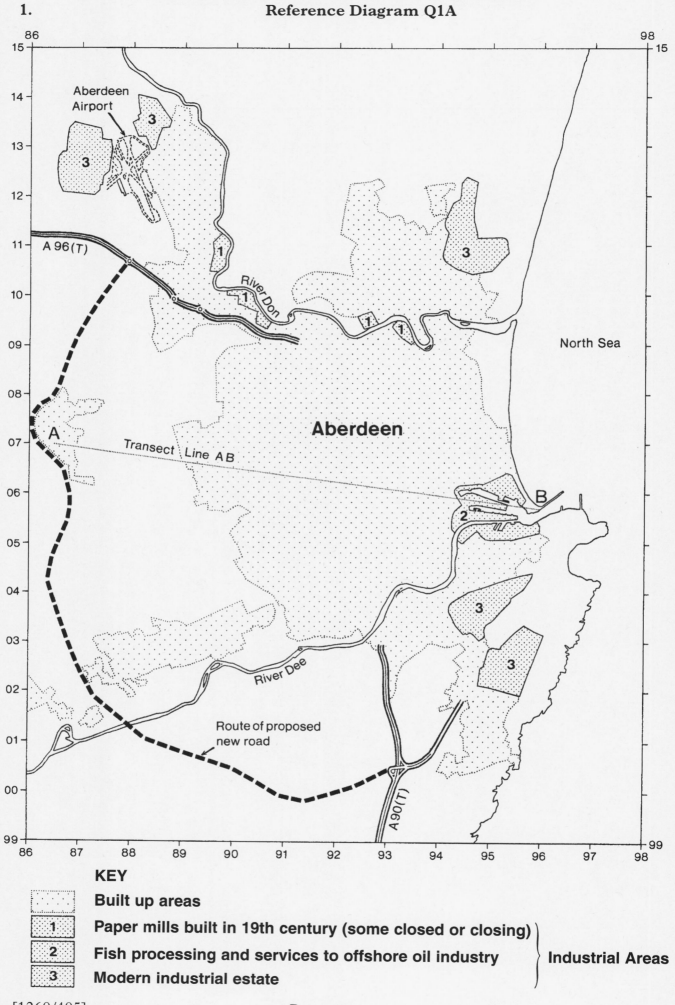

KEY
- Built up areas
- 1 Paper mills built in 19th century (some closed or closing)
- 2 Fish processing and services to offshore oil industry
- 3 Modern industrial estate

Industrial Areas

1. (continued)

This question refers to the OS Map Extract (No 1267/38) of Aberdeen and Reference Diagram Q1A on *Page two*.

(a) Look at Reference Diagram Q1A, showing the proposed route of a major new road.

Do you think this road should be built?

Use map evidence to justify your answer.

Reference Diagram Q1B: Model of Urban Land Use

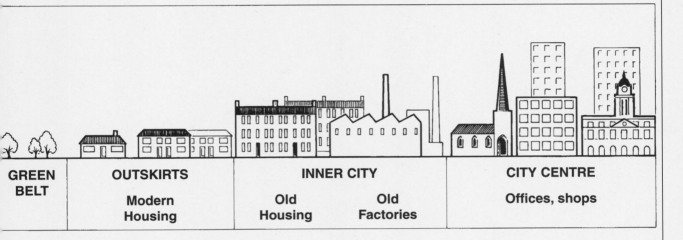

(b) Look at Reference Diagram Q1B above which shows a model of land use from the centre to the edge of a city.

Look at the OS map extract. Find the transect drawn from A (865070) to B (960057) on Reference Diagram Q1A.

Describe the similarities **and** differences in land use between the model and those found along transect AB.

(c) **Explain** the locations of the different industrial zones numbered **1**, **2** and **3** on Reference Diagram Q1A.

[Turn over

1. (continued)

(d) Study the course of the River Dee and its valley between 866010 and 929035. Describe how the **physical** features of the river and its valley have influenced land use.

(e) Air transport is increasingly important to the economy of cities such as Aberdeen.

Look at Aberdeen Airport shown on Reference Diagram Q1A and the Ordnance Survey Map extract.

Using map evidence, describe the advantages **and** disadvantages of the site of Aberdeen Airport.

2. **Reference Diagram Q2: A Corrie in the Scottish Highlands**

Look at Reference Diagram Q2.

Explain the formation of a corrie. You may use diagram(s) to illustrate your answer.

3. **Reference Diagram Q3: Synoptic Chart for British Isles, Sunday 10 December 2000**

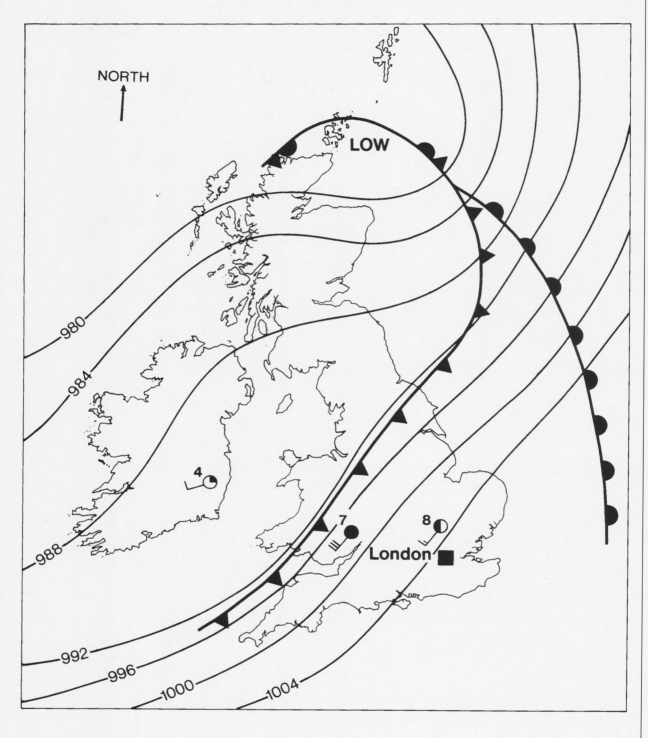

3. (continued)

Look at Reference Diagram Q3.

(a) Describe in detail the weather conditions being experienced in London on Sunday 10 December 2000. **4**

(b) "There will be some heavy showers of rain, giving way to sunny intervals. Winds, at first strong, will become lighter. Temperatures will drop."

(Forecast for London for Monday 11 December)

Give reasons for the changes which the weather forecasters believe will take place in London on Monday 11 December. **4**

[Turn over

4. Reference Diagram Q4A: Burgar Hill

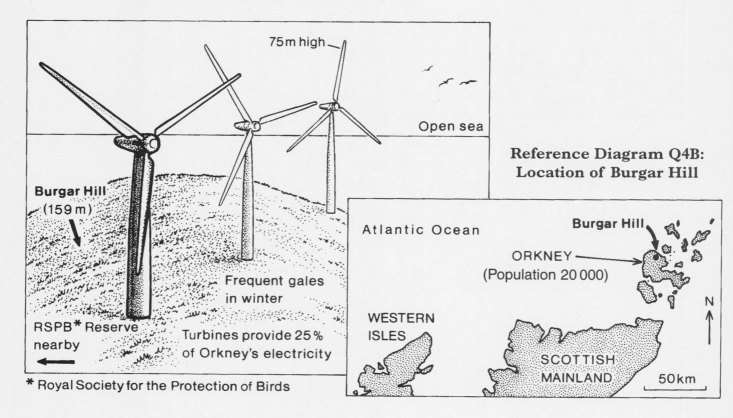

Look at Reference Diagrams Q4A and Q4B.

Three large wind turbines have been constructed on Burgar Hill in Orkney.

What are the advantages **and** disadvantages of having wind turbines in this area?

5. Reference Diagram Q5: Sphere of Influence of Inverness

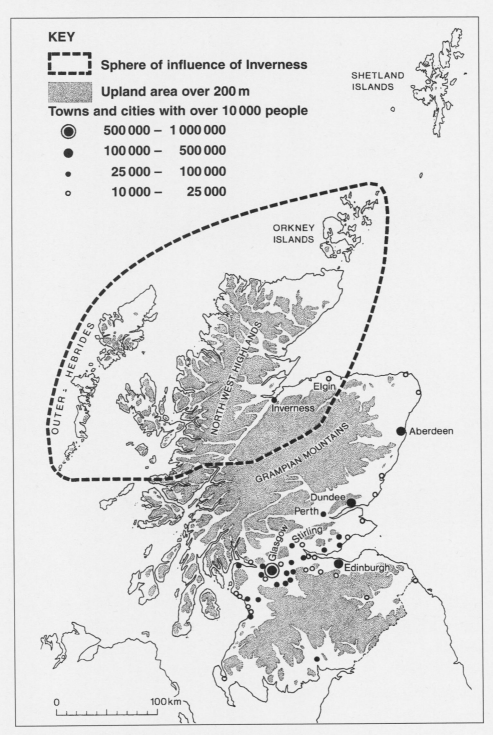

(a) Describe the gathering techniques which could have been used to identify the sphere of influence of Inverness as shown in Reference Diagram Q5.

Justify your choice of techniques.

(b) Look at Reference Diagram Q5.

The sphere of influence of Inverness has an unusual size and shape.

Suggest reasons for this.

6. Reference Diagram Q6A: Changing Land Use on Clook Farm

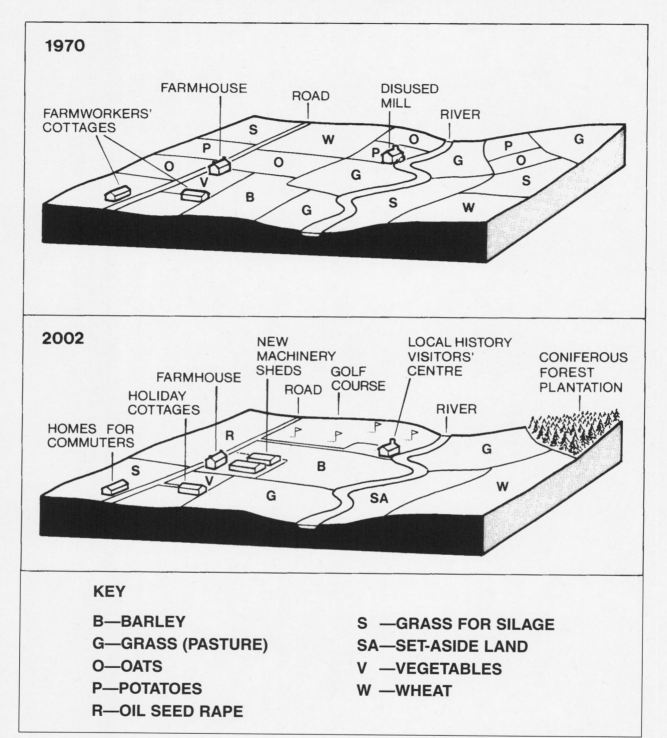

(a) Give reasons for the changes in land use on Clook Farm between 1970 and the present day.

6. (continued)

Reference Table Q6B: Percentage Land Use on Clook Farm

	1970	2002
Barley	10	15
Grass (pasture)	25	28
Oats	15	0
Potatoes	10	0
Oil seed rape	0	10
Grass for silage	15	12
Set aside land	0	10
Vegetables	10	5
Wheat	15	8
Golf course	0	7
Forestry	0	5

(b) Study Reference Table Q6B.

Suggest **two** other processing techniques which could be used to show this information. Give reasons for your choices.

5

[Turn over

7.

Reference Diagram Q7A:
Population Density of South Island, New Zealand

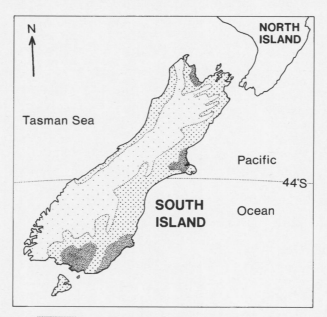

- Over 20 persons per square kilometre
- 2 to 20 persons per square kilometre
- 0 to 1 person per square kilometre

Reference Diagram Q7B:
Relief and Physical Resources

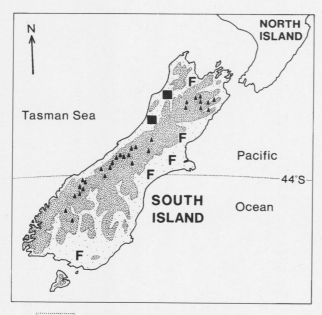

- ▲ Peaks over 2000 metres
- Land over 400 metres (The Southern Alps)
- Land under 400 metres
- F Best farmland
- ■ Coalfields

Reference Diagram Q7C:
Annual Rainfall

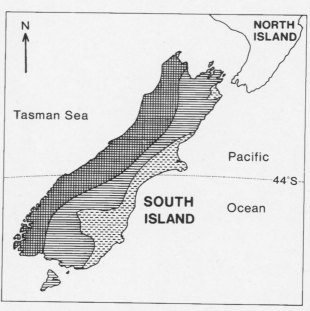

- Over 1700 mm per year
- 700 to 1700 mm per year
- Less than 700 mm per year

0 ———— 500 km

7. (continued)

(a) Look at Reference Diagrams Q7A, Q7B and Q7C.

Explain the population distribution on South Island, New Zealand.

4

[Turn over

7. (continued)

Reference Diagram Q7D: Population Pyramids for New Zealand and Indonesia

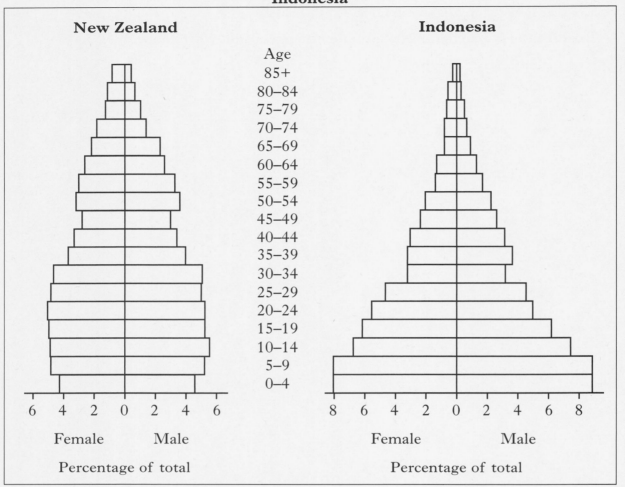

(b) Look at Reference Diagram Q7D.

"New Zealand is a developed country and Indonesia is a developing country."

Give reasons for the differences between the population structures of New Zealand and Indonesia.

8.   Reference Diagram Q8: World Trade

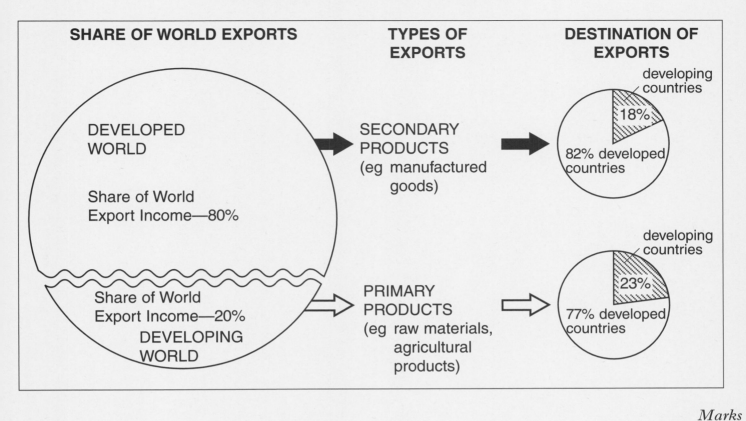

"The pattern of world trade benefits only countries of the Developed World."

Do you agree with this statement?

Give reasons for your answer.

5

[END OF QUESTION PAPER]

2003 | General

1260/403

NATIONAL QUALIFICATIONS 2003

THURSDAY, 15 MAY 10.25 AM–11.50 AM

GEOGRAPHY
STANDARD GRADE
General Level

Fill in these boxes and read what is printed below.

Full name of centre

Town

Forename(s)

Surname

Date of birth
Day Month Year

Scottish candidate number

Number of seat

1. Read the whole of each question carefully before you answer it.
2. Write in the spaces provided.
3. Where boxes like this ☐ are provided, put a tick ✓ in the box beside the answer you think is correct.
4. Try all the questions.
5. Do not give up the first time you get stuck: you may be able to answer later questions.
6. Extra paper may be obtained from the invigilator, if required.
7. Before leaving the examination room you must give this book to the invigilator. If you do not, you may lose all the marks for this paper.

Official SQA Past Papers: General Geography 2003

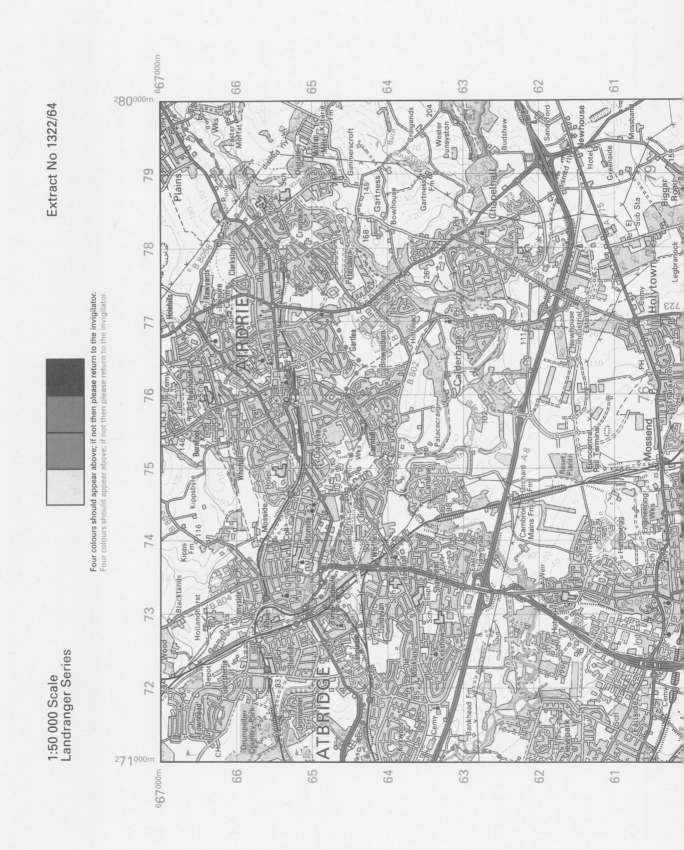

1:50 000 Scale
Landranger Series

Extract No 1322/64

Four colours should appear above; if not then please return to the invigilator.
Four colours should appear above; if not then please return to the invigilator.

Official SQA Past Papers: General Geography 2003

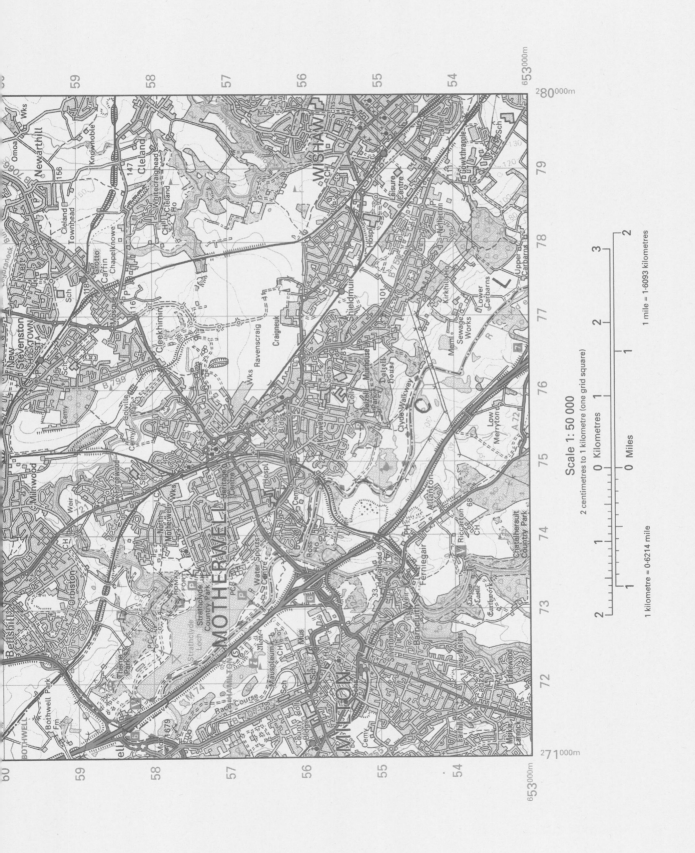

1. **Reference Diagram Q1A**

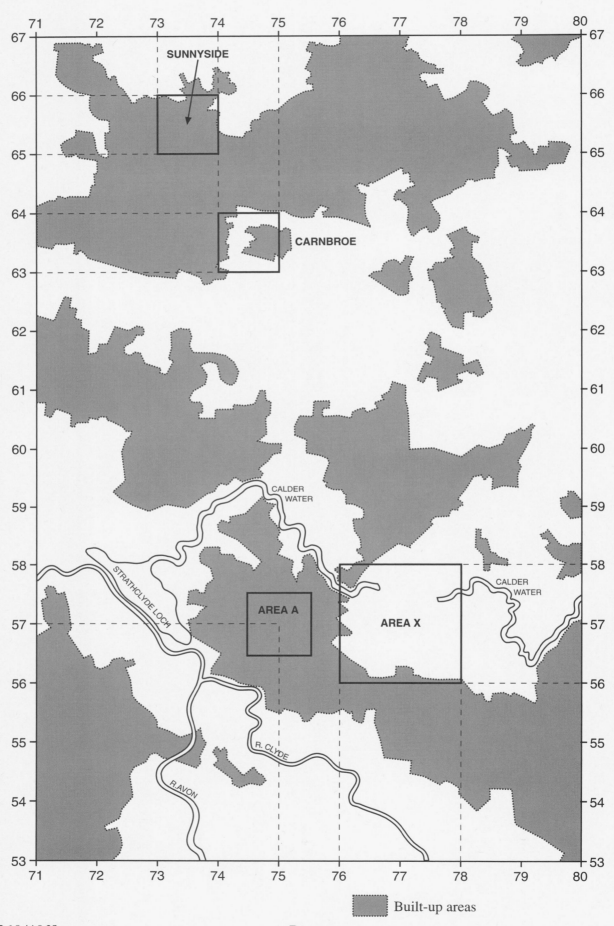

1. (continued)

Look at the Ordnance Survey Map Extract (No 1322/64) and at Reference Diagram Q1A on *Page two*.

(a) Give map evidence to show that the CBD of Motherwell is in Area A.

_____

_____

_____

3

(b) Two residential areas of Coatbridge are found in squares 7365 (Sunnyside) and 7463 (Carnbroe).

Describe the differences between these areas, referring to map evidence.

_____

_____

_____

_____

_____

4

(c) Find Bankhead Farm at 713630.

Using map evidence, describe the advantages **and** disadvantages of its location.

_____

_____

_____

_____

_____

4

[Turn over

1. (continued)

Reference Diagram Q1B: An old Ordnance Survey Map of part of Motherwell (1950 edition)

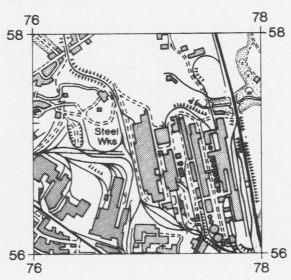

(d) (i) The area shown on the diagram above is identified as Area X on Reference Diagram Q1A.

Look at the Ordnance Survey Extract **and** the old Ordnance Survey Map above.

Describe the changes which have taken place between 1950 and the present day.

3

(ii) Comparing old and new maps is one technique for gathering data on land use.

State **two other** techniques that local pupils could use to gather information about land use change in this industrial area.

Give reasons for your choice.

Technique one _____

Technique two _____

Reasons _____

4

1. (continued)

(e) Strathclyde Country Park is centred on Strathclyde Loch.

Using map evidence, describe the attractions which this park has for visitors.

(f) Describe the **physical** features of the River Clyde **and** its valley between 774530 and 737560.

2. **Reference Diagram Q2: Landscapes of the Tay Valley**

Q2A: Upper Course of the River Tay

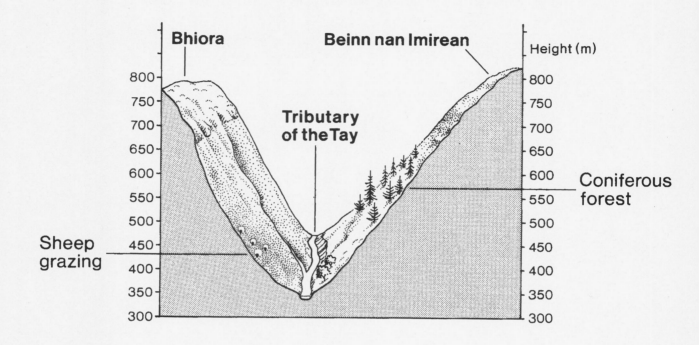

Q2B: Lower Course of the River Tay

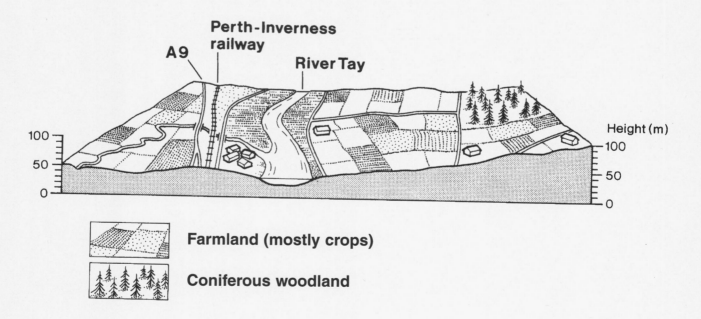

DO NOT WRITE IN THIS MARGIN

KU | ES

Marks

2. (continued)

(*a*) Look at the landscapes shown in the Reference Diagram opposite.

Compare the **physical** features of the River Tay and its valley in the two diagrams.

4

(*b*) **Explain** why land use along the River Tay is different in the two diagrams.

4

[Turn over

3. **Reference Diagram Q3A: Features of a Stevenson Screen**

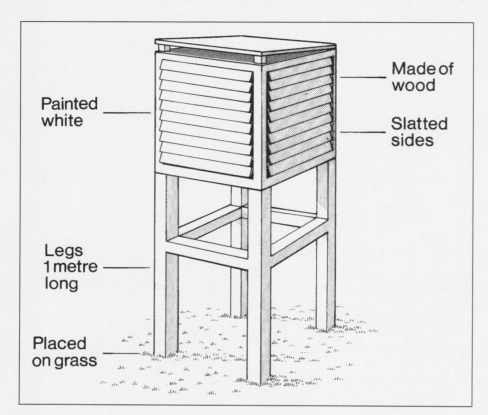

(a) Look at Reference Diagram Q3A which shows some design features of a Stevenson Screen which is used to house thermometers.

Choose **three** of these features and for each **explain** why it is necessary.

First Feature Chosen _____

Explanation _____
_____

Second Feature Chosen _____

Explanation _____
_____

Third Feature Chosen _____

Explanation _____
_____

3. (continued)

Reference Diagram Q3B: Weather Map of Mainland Scotland on 25 November 2001

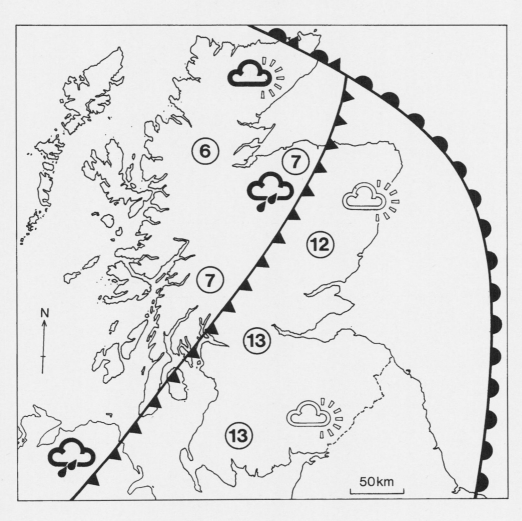

(b) Look at Reference Diagram Q3B.

Give reasons for the variations in temperatures throughout Scotland on 25 November.

4. Reference Table Q4A: Climate Statistics for Belem, Brazil

	J	F	M	A	M	J	J	A	S	O	N	D
Temperature (°C)	27	26	26	26	26	26	26	26	27	27	27	27
Rainfall (mm)	320	360	360	320	260	170	150	110	90	80	70	160

(a) Look at Reference Table Q4A.

Complete the rainfall graph for Belem on the grid below.

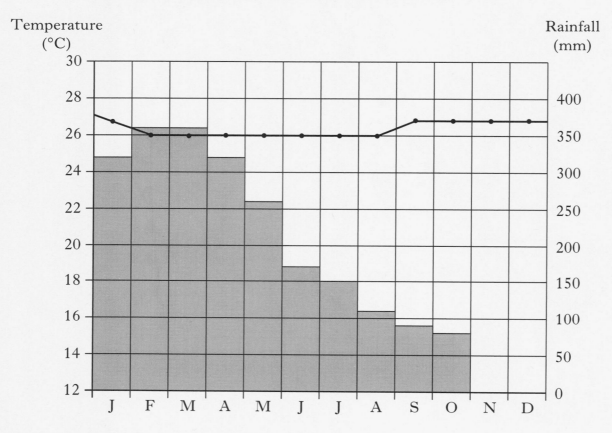

(b) Describe in detail the climate of Belem.

4. (continued)

Reference Diagram Q4B: Causes of Deforestation in Brazil

Trans Amazon Highway

Cattle Ranching

Open-Cast Iron Ore Mine

(c) Look at Reference Diagram Q4B above.

Describe the effects of the activities shown in the diagram on the environment **and** people of Brazil's rainforest.

4

5. Reference Diagram Q5A: Needs for a Modern Port

- Flat land
- Deep water
- Shelter
- Close to city

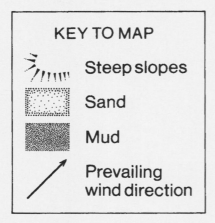

Reference Diagram Q5B: Potential Sites for a Modern Port

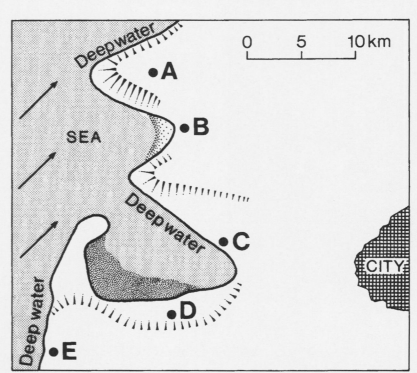

Look at Reference Diagrams Q5A and Q5B.

Which site—**A**, **B**, **C**, **D** or **E**—is the best for a modern port?

Choice _____

Give reasons for your answer.

_____
_____
_____
_____
_____

4

6. **Reference Diagram Q6A: Migration from Rural Areas in Developing Countries**

(a) Look at Reference Diagram Q6A.

People living in rural areas in developing countries can face many problems which may encourage them to migrate to cities.

Describe the type of problems found in such rural areas.

4

[Turn over

6. (continued)

Reference Diagram Q6B: Migrants' View of Life in a Developing City

(b) Look at Reference Diagram Q6B.

Do you think people benefit by moving from the countryside to the city? Explain your answer.

7. Reference Diagram Q7: Headquarters of the World's 100 Largest Companies

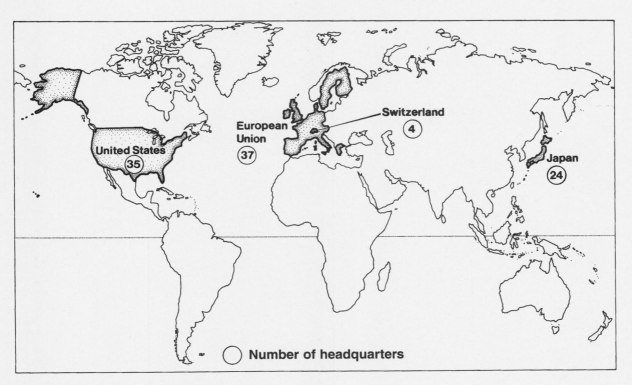

Look at Reference Diagram Q7 above.

(a) Give reasons for the location of the world's 100 largest companies.

(b) Give **one other** technique that could be used to process the information shown on the map.

Give reasons for your choice.

8. **Reference Diagram Q8: Aid to Developing Countries**

Short-term Aid — Immediate help

Long-term Aid — Helps a country to develop

- Clean water
- Food
- Emergency shelter
- Medicines

- Rebuilding homes
- Road building
- Electricity network
- Building hospitals

Look at Reference Diagram Q8.

Which type of aid, short-term or long-term, would be most useful to a **developing** country after an earthquake?

Give reasons for your answer.

4

[END OF QUESTION PAPER]

2003 | Credit

[BLANK PAGE]

1260/405

NATIONAL QUALIFICATIONS 2003

THURSDAY, 15 MAY 1.00 PM – 3.00 PM

GEOGRAPHY STANDARD GRADE Credit Level

All questions should be attempted.

Candidates should read the questions carefully. Answers should be clearly expressed and relevant.

Credit will always be given for appropriate sketch-maps and diagrams.

Write legibly and neatly, and leave a space of about one cm between the lines.

Marks may be deducted for bad spelling and bad punctuation, and for writing that is difficult to read.

All maps and diagrams in this paper have been printed in black only: no other colours have been used.

Extract No 1323/36/43

1:50 000 Scale
Landranger Series

Four colours should appear above; if not then please return to the invigilator.
Four colours should appear above; if not then please return to the invigilator.

Official SQA Past Papers: Credit Geography 2003

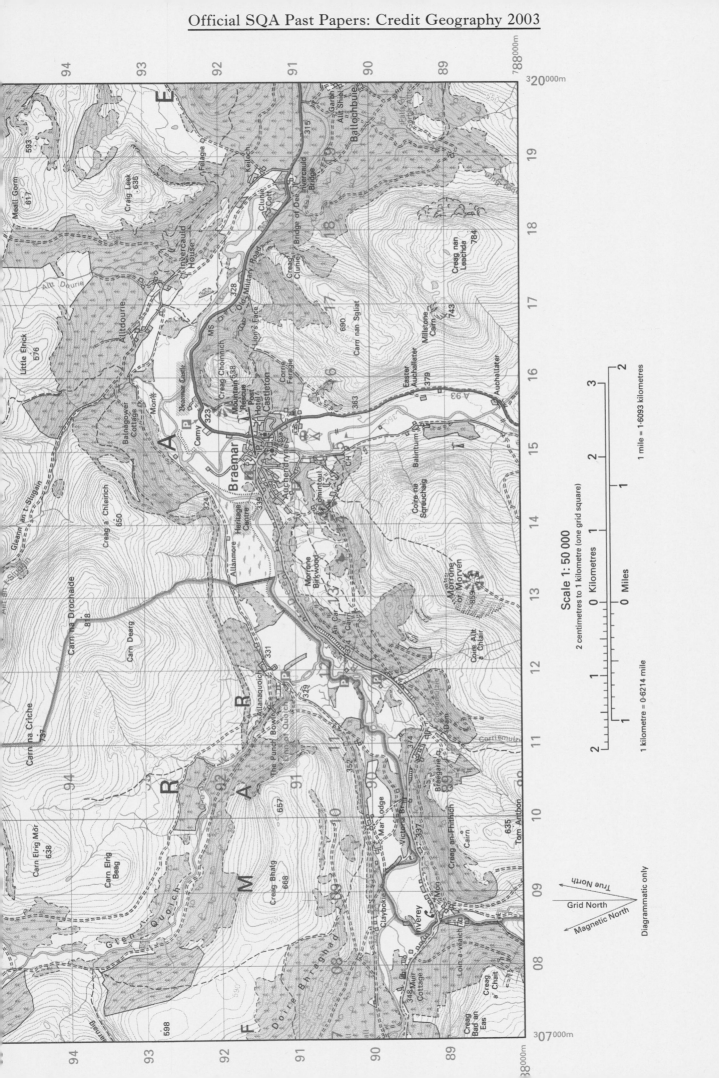

1. **Reference Diagram Q1A**

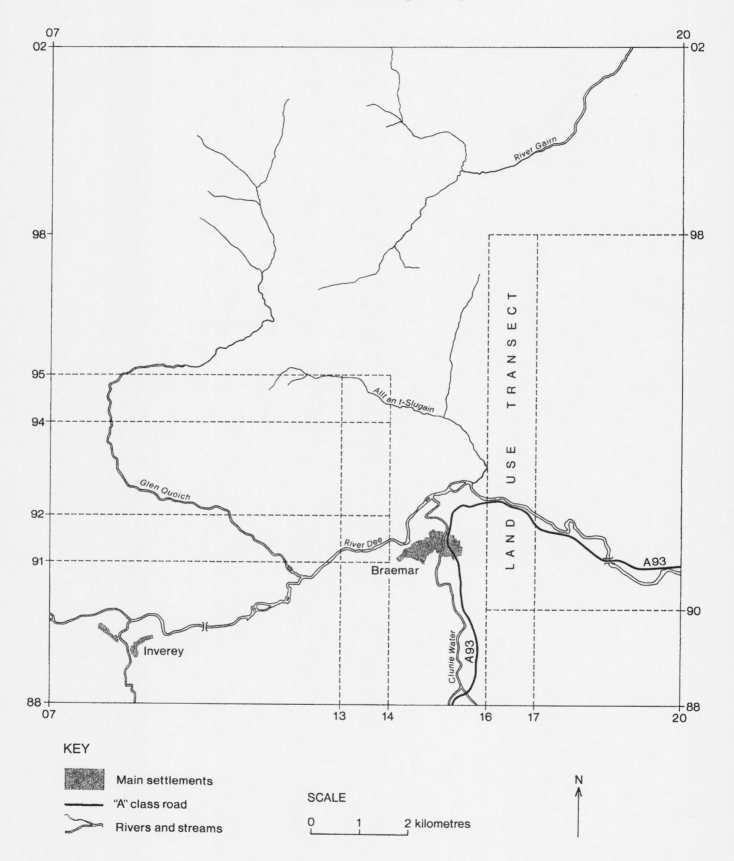

1. (continued)

This question refers to the OS Map Extract (No 1323/36/43) of the Braemar Area and the Reference Diagram Q1A on *Page two*.

(a) (i) Match each of the features named below with the correct grid reference.

Features: hanging valley; truncated spur; corrie; U shaped valley.

Choose from grid references: 094992, 134996, 155993, 146980.

3

(ii) **Explain** how **one** of these features listed in (a)(i) was formed.

You may use diagrams to illustrate your answer.

4

Reference Diagram Q1B: Requirements for National Park Status

High quality landscape   *Variety of plant and animal habitats*

Recreation and tourist value   *Historic features*

Reference Diagram Q1C: Main Land Uses in the Braemar Area

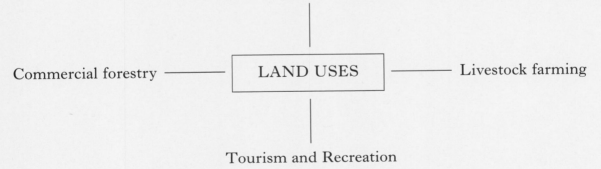

(b) Look at Reference Diagrams Q1B and Q1C.

The whole of the area covered by the OS map extract is within the recently designated Cairngorm National Park.

Describe the advantages **and** disadvantages this area has for a national park.

Use map evidence to support your answer.

6

[Turn over

1. (continued)

(c) Study Reference Diagram Q1A and the OS map extract.

The Rivers Dee (1391) and Allt an t-Slugain (1394) and their valleys are very different.

Describe these differences **in detail**.

Reference Diagram Q1D: Settlement Pattern around Braemar

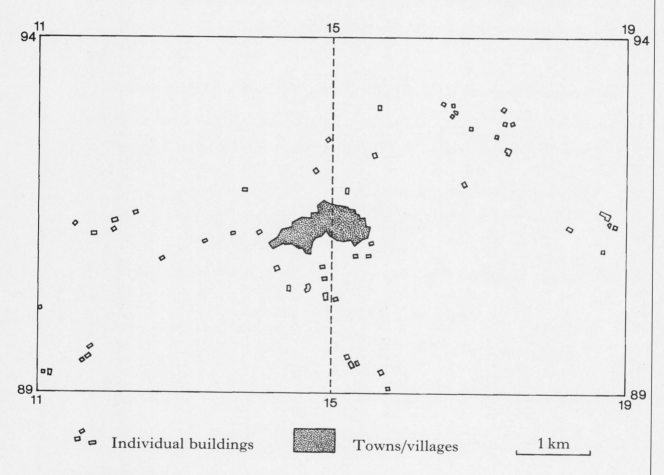

(d) **Explain** the distribution of settlement in the area of the map extract as shown by Reference Diagram Q1D.

1. (continued)

Reference Diagram Q1E: Land Use Transect

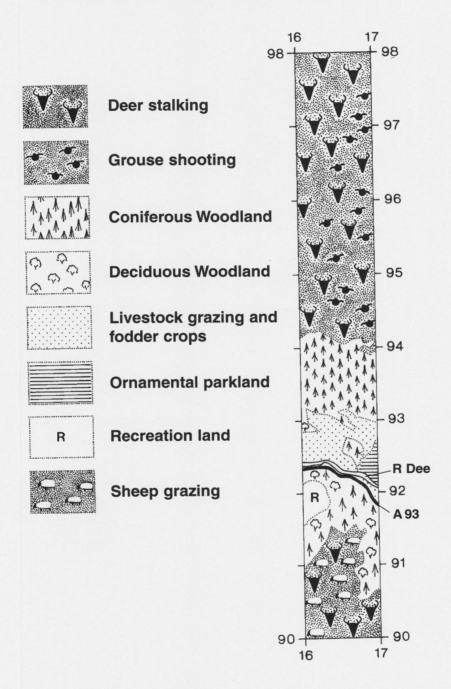

(e) Look at Reference Diagram Q1E, Reference Diagram Q1A and the OS map.

Give reasons for the pattern of land use which is shown on the transect.

6

2. Reference Diagram Q2A: European Synoptic Chart for Noon, 14 July 2001

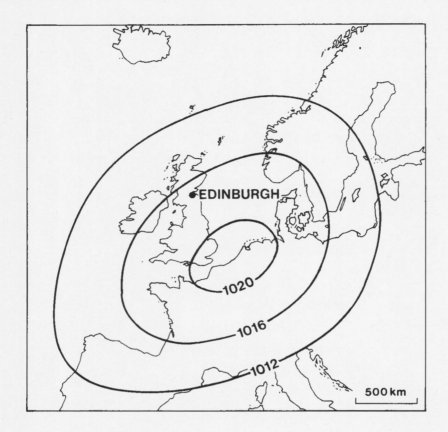

Reference Diagram Q2B: Three Weather Station Circles

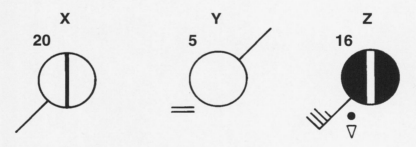

Study Reference Diagrams Q2A and Q2B.

State which weather station circle (**X**, **Y** or **Z**) shows the weather conditions at Edinburgh at noon on 14 July 2001.

Explain your choice **in detail**.

3. Reference Diagram Q3: Changing Landscape in West Africa

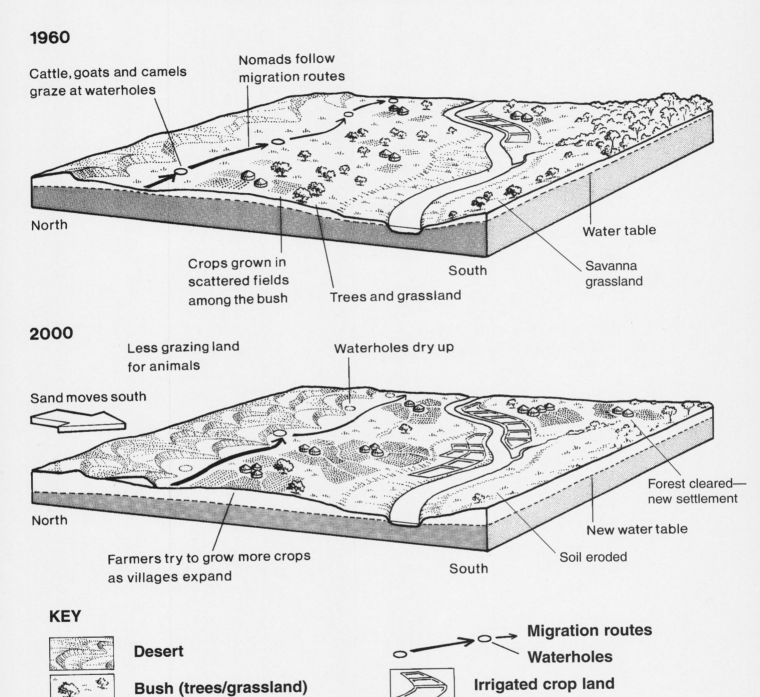

Look at Reference Diagram Q3.

Give reasons for the changes in the landscapes of West Africa.

6

4. **Reference Diagram Q4A: Site of Shrewsbury**

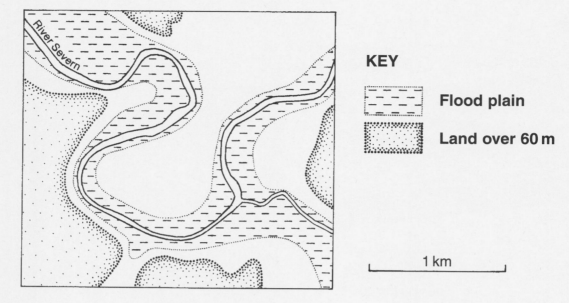

Reference Diagram Q4B:
Shrewsbury in 16th century

Reference Diagram Q4C:
Shrewsbury in 2000

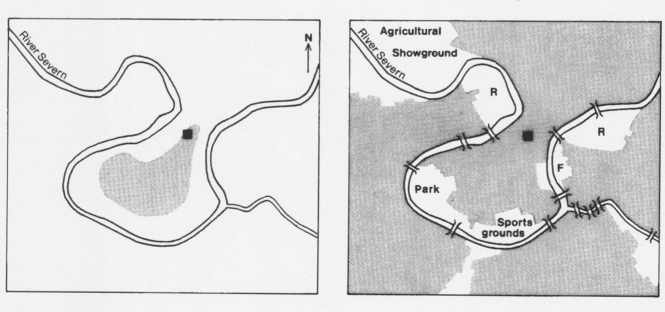

Look at Reference Diagrams Q4A, Q4B and Q4C.

Explain the ways in which the River Severn has influenced Shrewsbury in terms of its site, growth and land use.

5. Reference Diagram Q5: Sketch of Keilor Farm

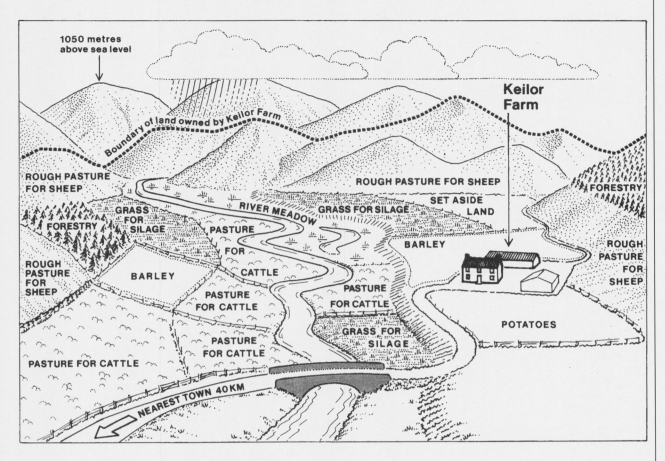

Study Reference Diagram Q5.

Keilor farm is a hill farm producing mainly beef cattle and sheep.

Explain the links between land use and the physical and human factors affecting the farm.

6

6. Reference Diagram Q6A: Nissan Car Factory, Washington (View looking South)

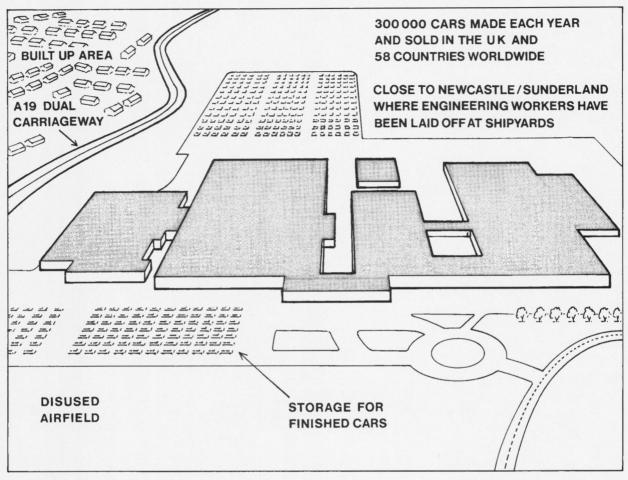

Reference Diagram Q6B: Location of Nissan Car Factory

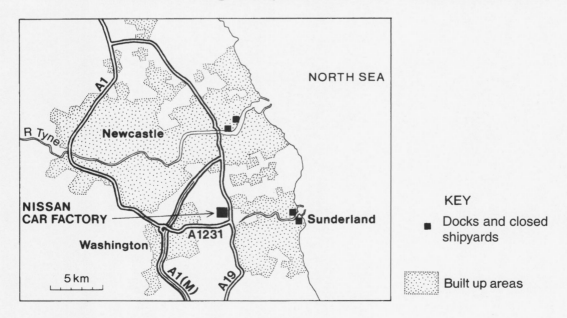

Study Reference Diagrams Q6A and Q6B.

Suggest reasons why Nissan chose to locate their car manufacturing plant at this site in Washington, NE England.

7. Reference Diagram Q7: The Millennium Link Canal Project

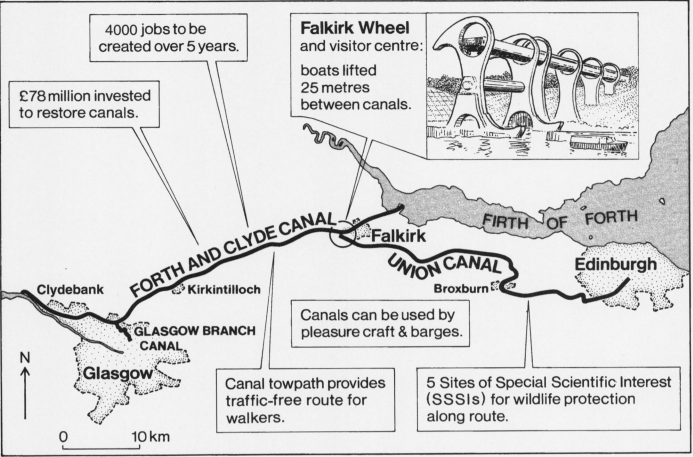

(a) Look at Reference Diagram Q7.

The Millennium Link has restored the Forth and Clyde Canal and the Union Canal in Central Scotland. It links the estuaries of the Rivers Forth and Clyde together as well as the cities of Edinburgh and Glasgow.

Describe the benefits which the opening of the Millennium Link will have for the economy and environment of the areas around it. **5**

(b) A group of geography students is researching the effects of the opening of the Millennium Link on the communities along its route.

Describe **two** gathering techniques they could use to do this.

Give reasons for your choices. **5**

[Turn over

8.

Reference Diagram Q8A: Location of Bolivia

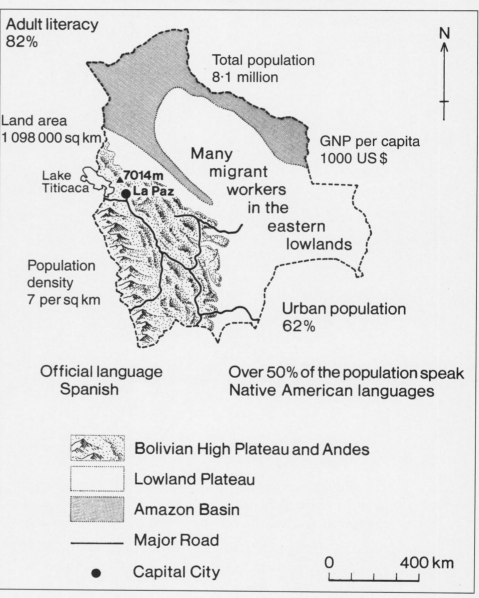

Look at Reference Diagrams Q8A and Q8B.

(a) Give reasons why it may be difficult to take an accurate census in **a developing** country such as Bolivia.

(b) What use could the Government of **a developing** country make of population census data?

9. **Reference Table Q9: Key Mineral Exports of South Africa**

Mineral	Percentage of World Reserves	World Rank	Percentage of World Production	World Rank
Chrome	68%	1	44%	1
Vermiculite	40%	2	46%	1
Gold	39%	1	21%	1
Titanium	31%	1	27%	2
Manganese	81%	1	14%	3
Uranium	7%	5	4%	9

Look at Reference Table Q9.

Identify other **techniques** that could be used to **process** the data shown above. Give reasons for your choices.

5

[END OF QUESTION PAPER]

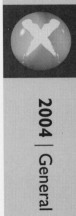

2004 | General

[BLANK PAGE]

FOR OFFICIAL USE

G

KU | ES

Total Marks

1260/403

NATIONAL QUALIFICATIONS 2004

MONDAY, 17 MAY 10.25 AM–11.50 AM

GEOGRAPHY STANDARD GRADE General Level

Fill in these boxes and read what is printed below.

Full name of centre

Town

Forename(s)

Surname

Date of birth
Day Month Year

Scottish candidate number

Number of seat

1 Read the whole of each question carefully before you answer it.

2 Write in the spaces provided.

3 Where boxes like this ☐ are provided, put a tick ✓ in the box beside the answer you think is correct.

4 Try all the questions.

5 Do not give up the first time you get stuck: you may be able to answer later questions.

6 Extra paper may be obtained from the invigilator, if required.

7 Before leaving the examination room you must give this book to the invigilator. If you do not, you may lose all the marks for this paper.

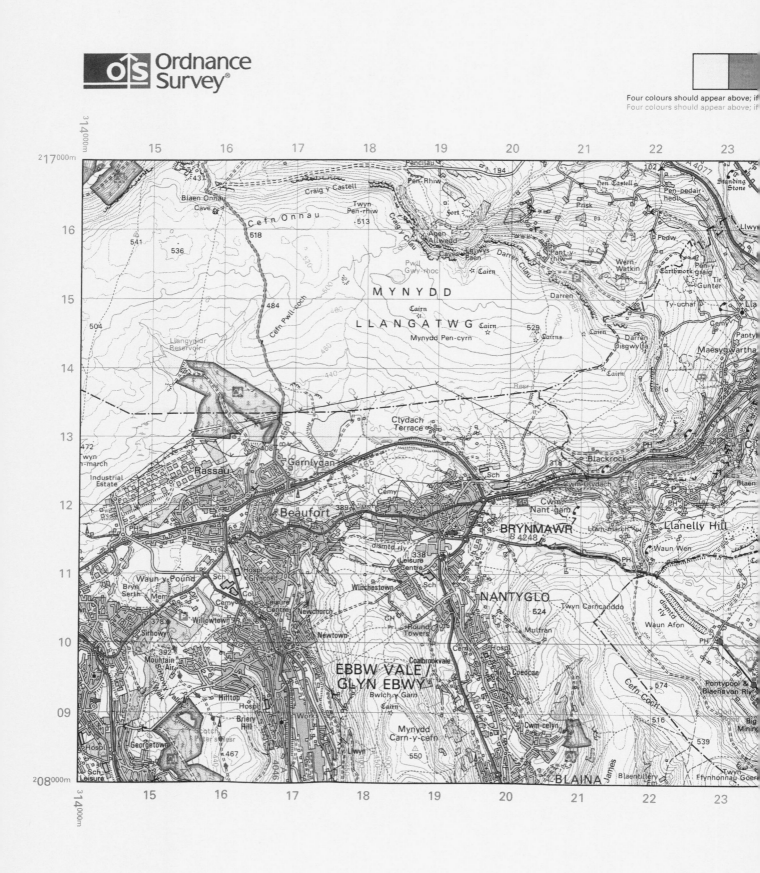

Official SQA Past Papers: General Geography 2004

Extract No 1348/161

1:50 000 Scale
Landranger Series

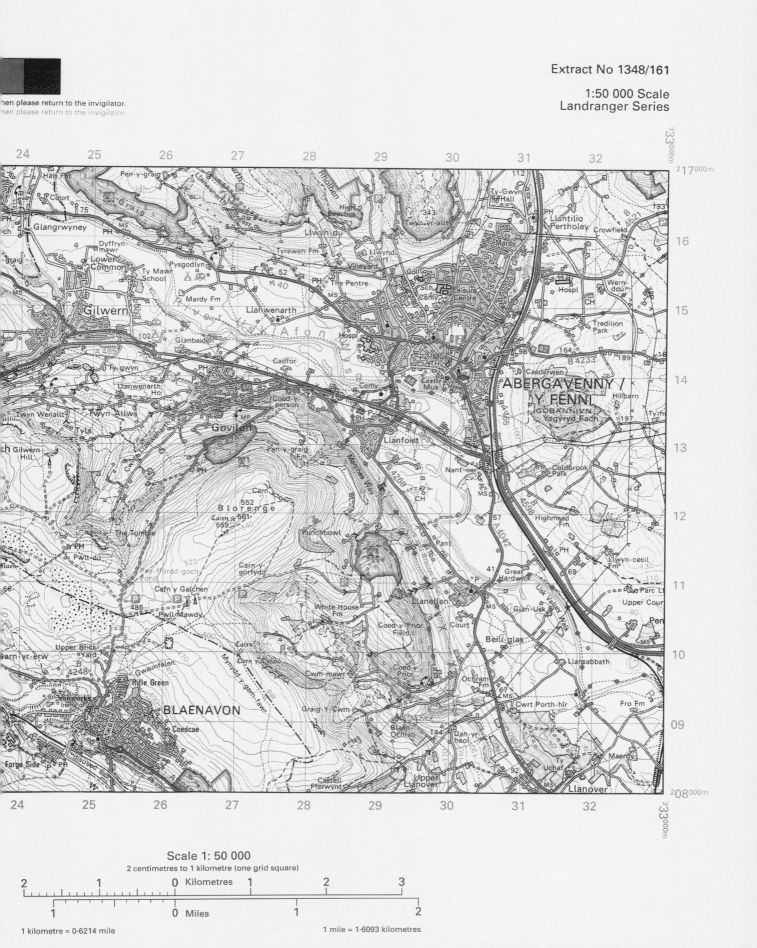

Scale 1: 50 000
2 centimetres to 1 kilometre (one grid square)

1 kilometre = 0·6214 mile   1 mile = 1·6093 kilometres

1. **Reference Diagram Q1A**

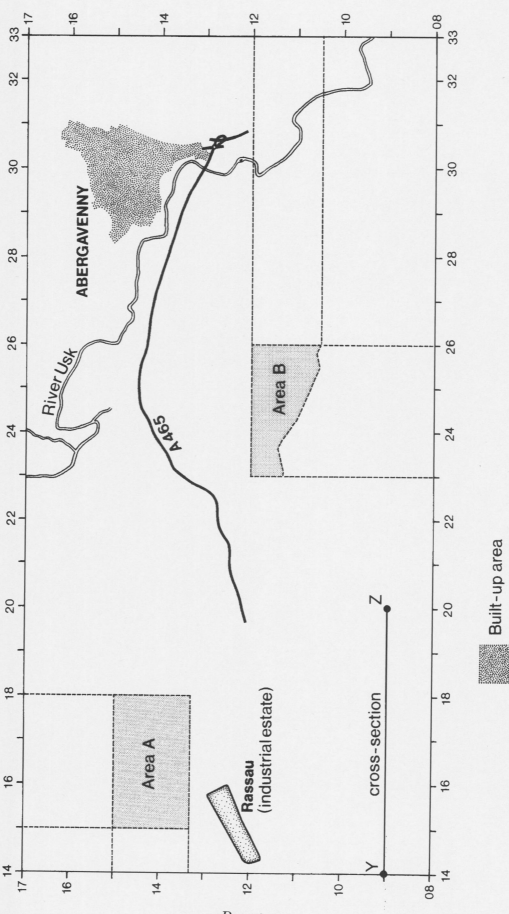

1. (continued)

Look at the Ordnance Survey Map Extract (No 1348/161) of the Ebbw Vale/Abergavenny area and Reference Diagram Q1A on *Page two*.

(*a*) Complete the table below by matching the physical features to the correct grid references.

Choose from the following grid squares.

    1814   2316   2112   1808

Physical Feature	Grid Square
Deep narrow valley	
Ridge between two valleys, over 500 metres	
Broad flood plain	
Part of gentle slope, facing south west	

3

[Turn over

1. (continued)

Reference Diagram Q1B: Cross-section YZ from 140090 to 200090

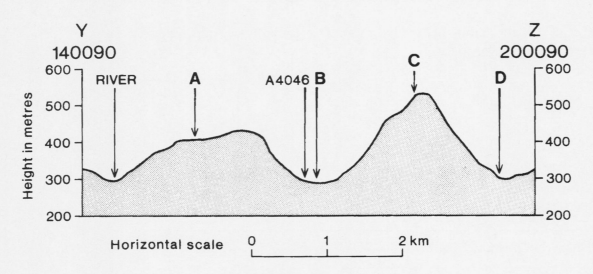

(b) Look at Reference Diagram Q1B. Find this cross-section on the Ordnance Survey map.

Match the features (A, B, C and D) on the cross-section YZ with the correct descriptions in the table below.

Feature	Letter
Works	
A467	
Cairn	
Scotch Peter's Reservoir	

3

(c) In what ways has the **physical** landscape created problems for engineers in building the A465 road from grid square 1912 to grid square 3012?

_____

_____

_____

_____

_____

4

1. (continued)

(d) What is the main function of the town of Abergavenny?

Tick (✓) your choice.

Tourist resort ☐   Market town ☐

Give reasons for your choice.

_____
_____
_____
_____
_____

4

(e) **Reference Diagram Q1C: Selected Aims of National Parks**

> * preserve the beauty of the countryside
> * conserve the local wildlife
> * provide good access and facilities for public open air enjoyment
> * maintain established farming

Areas A and B on Reference Diagram Q1A are in the Brecon Beacons National Park. Find them on the map extract.

For each area, **explain** how land use is in conflict with the aims shown above.

Area A _____
_____
_____

Area B _____
_____
_____

4

[Turn over

1. **(continued)**

   (f) There is an industrial estate at Rassau, grid squares 1412/1512.

   What are the advantages **and** disadvantages of this location for an industrial estate?

   Advantages _____

   _____

   _____

   _____

   Disadvantages _____

   _____

   _____

   _____

   4

2. **Reference Diagram Q2: How a Waterfall develops**

STAGE 1

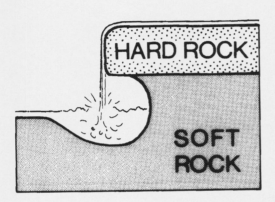

STAGE 2

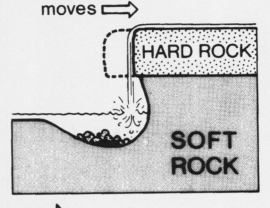

Change over time

Study Reference Diagram Q2.

Explain, in detail, why a waterfall moves upstream from its original position.

[Turn over

3. Reference Diagram Q3: Synoptic Chart, 12 December 0600 hours

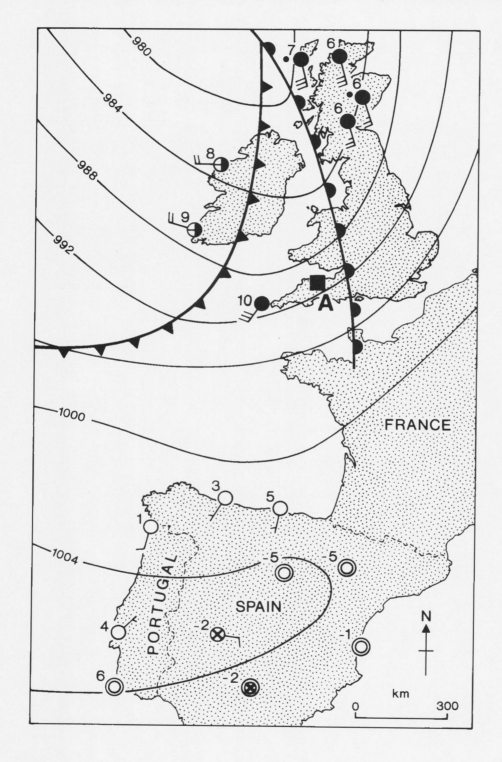

3. (continued)

(a) Complete the station circle below to show the weather conditions at **A** on Reference Diagram Q3.

Weather conditions at A
Wind from South West
Cloud cover: 7 oktas
Rain
Wind speed: 15 knots

3

(b) Study Reference Diagram Q3.

Match the weather systems to the locations given in the table.

Choose from: Anticyclone   Depression

Location	Weather System
British Isles	
Spain	

Give reasons for your answer.

_____
_____
_____
_____
_____

4

[Turn over

4. **Reference Diagram Q4A: Tropical Rainforest Climate**

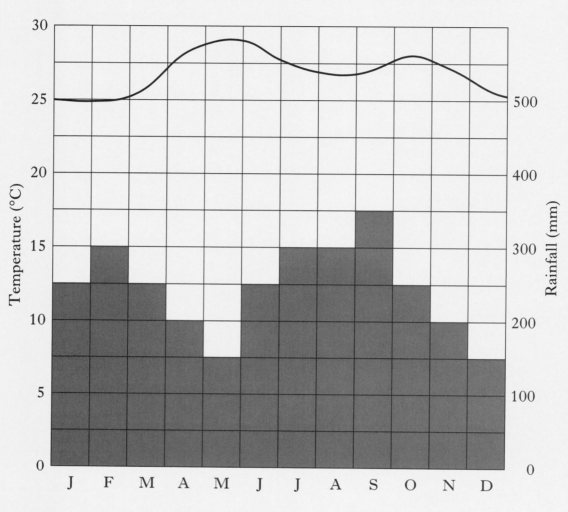

(a) Look at Reference Diagram Q4A.

Describe, **in detail**, the climate of the tropical rainforest.

4. (continued)

Reference Diagram Q4B: Tropical Rainforest Landscape Before Deforestation

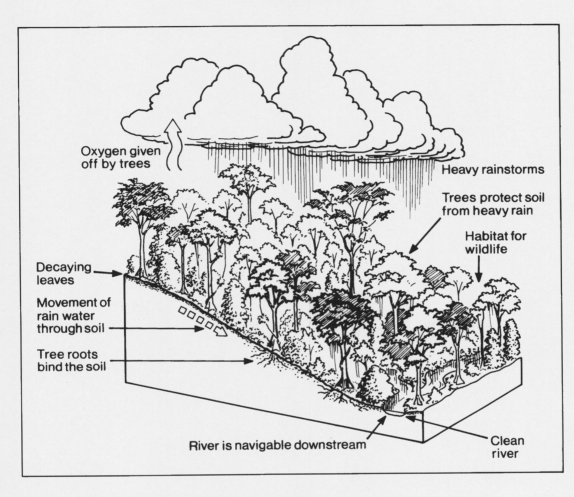

(b) Look at Reference Diagram Q4B.

Explain problems caused by deforestation in the Tropical Rainforest.

5. Reference Diagram Q5: Relief, Climate and Selected Land Uses

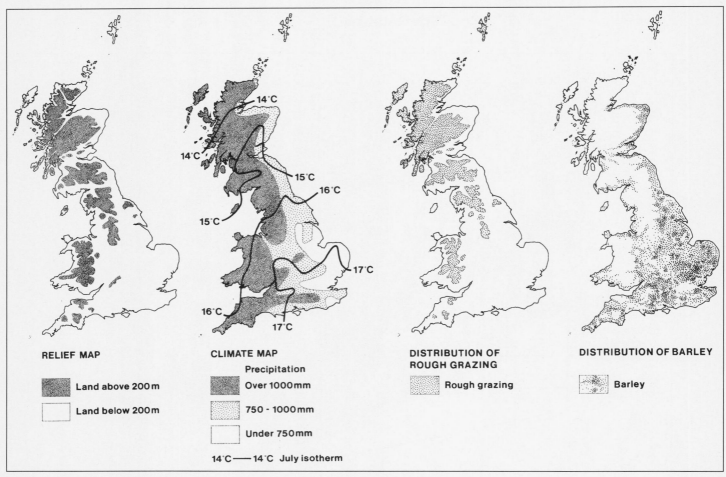

Look at Reference Diagram Q5.

What influence do relief and climate have on the distribution of rough grazing and barley?

[Turn over for Question 6 on *Page fourteen*

6. **Reference Diagram Q6A: The Inverfirth Estuary in 1974**

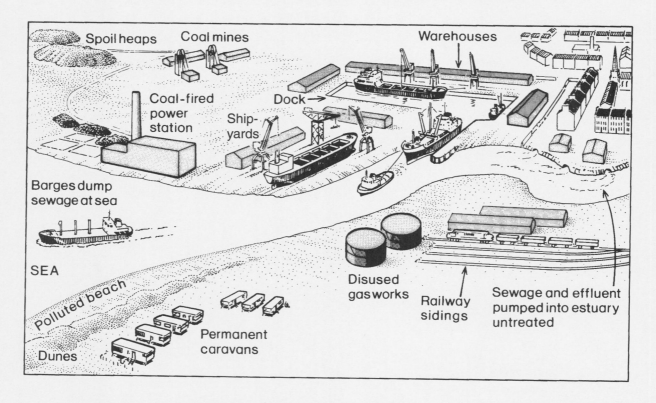

Reference Diagram Q6B: The Inverfirth Estuary in 2004

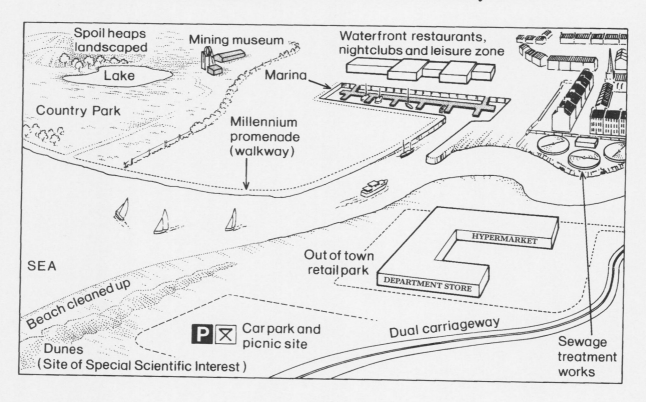

6. (continued)

(a) Study Reference Diagrams Q6A and Q6B.

What techniques could pupils have used to gather the information shown on Reference Diagrams Q6A and Q6B?

Give reasons for your choice of techniques.

Techniques _____

_____

_____

Reasons _____

_____

_____

_____

4

(b) Do you think the changes that have taken place between 1974 and 2004 have improved the area?

Give reasons for your answer.

_____

_____

_____

_____

_____

4

[Turn over

7. Reference Diagram Q7A: Shopping Centres in a large Town

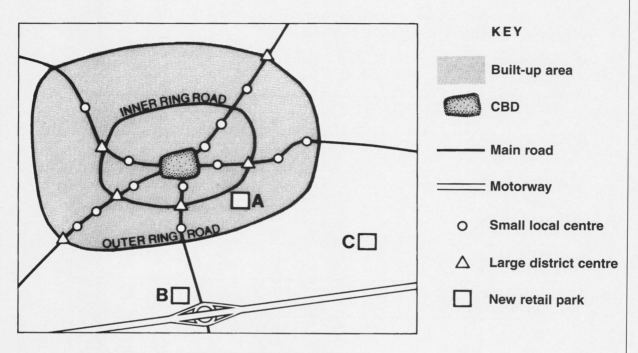

(a) Look at Reference Diagram Q7A above.

A, B and C are proposed sites for the development of a large retail park.

Which site do you think is best for this development? Give reasons for your choice.

Site _____

Reasons _____

_____

_____

_____

_____

4

7. (continued)

Reference Diagram Q7B: Employment Changes in the UK

		1980	1990	2000
Sectors of Industry (numbers in millions)	Primary	0·6	0·6	0·5
	Secondary	6·7	5·3	4·3
	Tertiary	16·8	20·1	22·3
	Total	24·1	26·0	27·1

(b) Give **two** processing techniques which could be used to present the information shown in Reference Diagram Q7B.

Give reasons for your choices.

Technique 1 _____

Reason _____

_____

_____

Technique 2 _____

Reason _____

_____

_____

4

[Turn over

8. Reference Diagram Q8: Population Pyramids

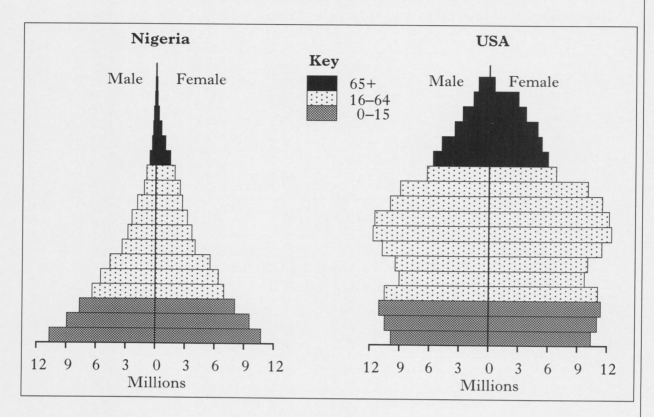

(a) Look at Reference Diagram Q8.

Describe three differences between the population pyramids of Nigeria and the United States of America.

3

8. (continued)

(b) (i) **Describe** two problems which the population structure of Nigeria may cause.

_____

_____

_____

_____

(ii) **Describe** two problems which the population structure of the United States may cause.

_____

_____

_____

_____

4

[Turn over for Question 9 on *Page twenty*

9. Reference Diagram Q9A: **Factors Affecting Sugar Production in the EU**

* high subsidies to farmers to produce beet sugar in the EU
* EU buys sugar in bulk from Mozambique
* import tariffs protect EU farmers from foreign competition
* good climate for growing sugar beet

Reference Diagram Q9B: **Factors Affecting Sugar Production in Mozambique**

* lowest production costs in the world
* Mozambique has to sell sugar to EU at low prices
* cannot sell processed (refined) sugar to EU because of high import tariffs on processed food
* good growing conditions for sugar cane

Give reasons why sugar producers in Mozambique are at a **disadvantage** compared with sugar producers in the EU.

4

[END OF QUESTION PAPER]

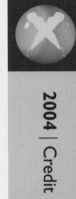

2004 | Credit

1260/405

NATIONAL QUALIFICATIONS 2004

MONDAY, 17 MAY 1.00 PM – 3.00 PM

GEOGRAPHY
STANDARD GRADE
Credit Level

All questions should be attempted.

Candidates should read the questions carefully. Answers should be clearly expressed and relevant.

Credit will always be given for appropriate sketch-maps and diagrams.

Write legibly and neatly, and leave a space of about one cm between the lines.

Marks may be deducted for bad spelling and bad punctuation, and for writing that is difficult to read.

All maps and diagrams in this paper have been printed in black only: no other colours have been used.

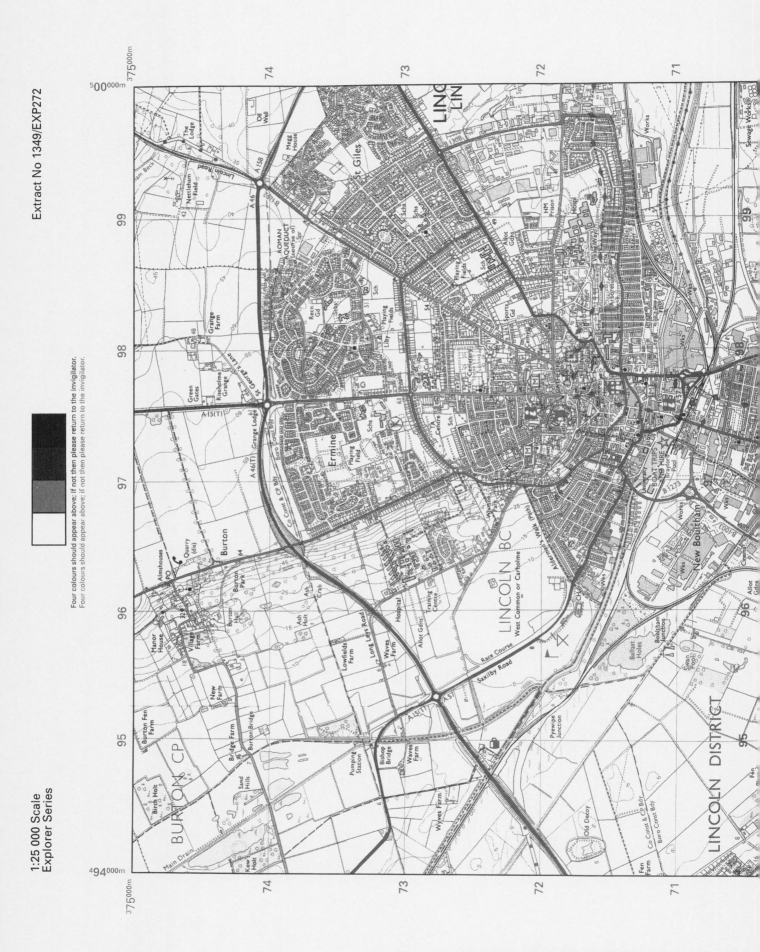

Official SQA Past Papers: Credit Geography 2004

1. **Reference Diagram Q1A**

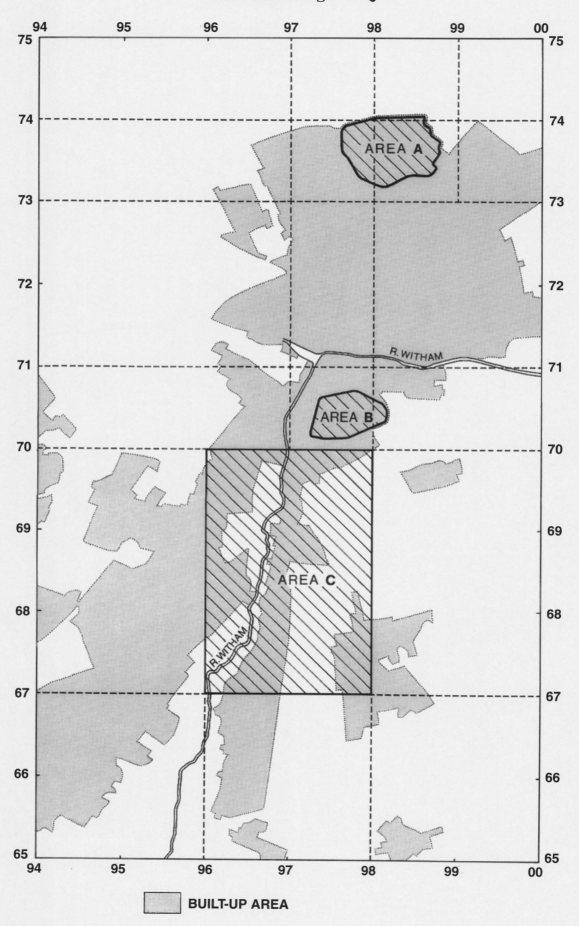

BUILT-UP AREA

1. (continued)

This question refers to the OS Map Extract (No 1349/EXP272) of Lincoln and the Reference Diagram Q1A on *Page two*.

(a) Give the Grid Reference of the square which contains the CBD of Lincoln. Support your answer with map evidence.

3

Look at Reference Diagram Q1A.

(b) Referring to map evidence, describe the differences between the residential environments of Area A and Area B.

4

(c)
> "A dormitory settlement is a community where most of the residents travel to work in a larger settlement."

Pupils from a local high school want to find out if Bracebridge Heath (9767–9867) is a dormitory settlement for Lincoln. What techniques could they use to gather relevant information?

Explain the choice of techniques.

5

(d) Give reasons for the differences between the leisure activities located in square 9771 and those located in squares 9468 and 9469.

4

(e) Suggest the type of farming found at Canwick Manor Farm (993677).

Give reasons for your choice.

4

(f) Referring to map evidence, explain how physical landscape features (relief and drainage) have affected land use in Area C.

6

[Turn over

2. Reference Diagram Q2A: Glaciated Upland

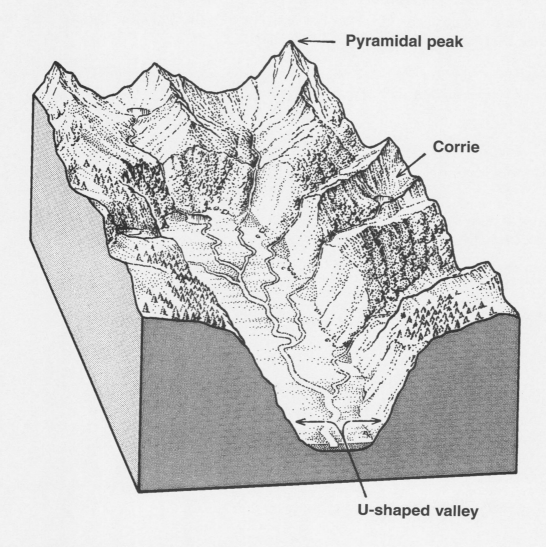

(a) Study Reference Diagram Q2A.

Select **one** of the labelled glaciated features shown.

Explain in detail how it was formed. You may wish to use diagrams to illustrate your answer.

2. (continued)

Reference Diagram Q2B: Some Land Uses in Glaciated Uplands

> Forestry
> Tourist Resorts
> Skiing
> Hydroelectric Power
> Farming

(b) Select **one** land use shown in Reference Diagram Q2B.

Explain in detail why the land use is suited to a glaciated upland as shown in Reference Diagram Q2A.

4

[Turn over

3. **Reference Diagram Q3: Synoptic Chart for 25 September 2002**

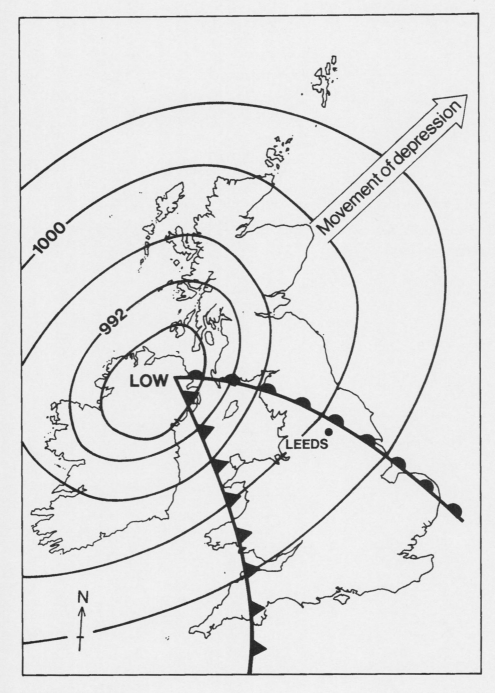

Study Reference Diagram Q3.

Explain the changes which will take place in the weather at Leeds in the next 12 hours.

4. **Reference Diagram Q4: Threats to the Marine Environment around Scotland**

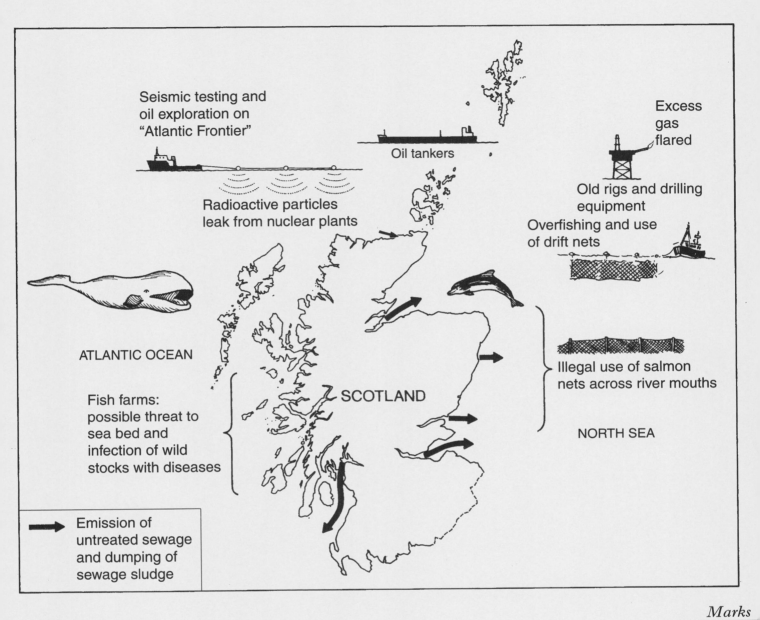

What measures could be taken to reduce the impact of the threats to the marine environment as shown on Reference Diagram Q4?

5. **Reference Diagram Q5A: Physical Data for two Farms in Scotland**

	Farm A	Farm B
Altitude	220 m to 450 m	75 m to 125 m
Average rainfall per year	1520 mm	630 mm
Sunshine hours per year	1000	1300

Reference Diagram Q5B: Other Data for the two Farms

	Farm A	Farm B
Area	1904 ha	444 ha
Workers	3 full time	5 full time 13 part time/seasonal
Machinery	2 tractors 8 other machines	6 tractors 14 other machines
Land use	80% sheep grazing 13% beef cattle grazing 7% barley, turnips and hay	87% arable—mainly wheat and barley with some potatoes, raspberries and strawberries 13% beef cattle grazing

(a) Look at Reference Diagrams Q5A and Q5B.

Give reasons for the differences between the two farms.

(b) Describe other techniques which could be used to present the land use data shown in Reference Diagram Q5B.

Give reasons for your choice of techniques.

6. **Reference Diagram Q6A: Location of Toyota Car Factory at Burnaston**

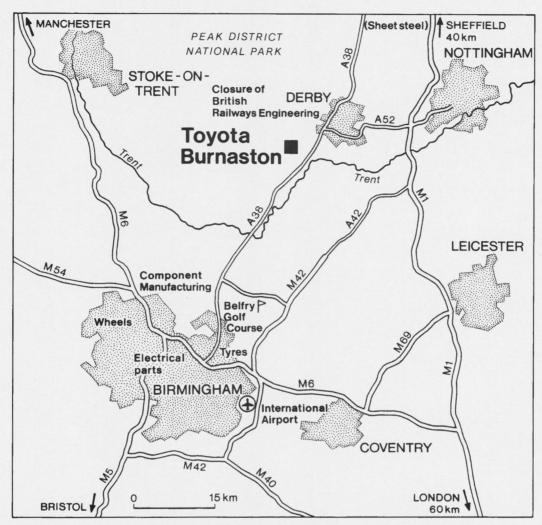

Reference Diagram Q6B: Site of Toyota's Burnaston Factory

Look at Reference Diagrams Q6A and Q6B.

What are the advantages of locating a car factory at Burnaston?

6

7. Reference Diagram Q7A: Destination and Origins of Migrants into and within Europe since 1990

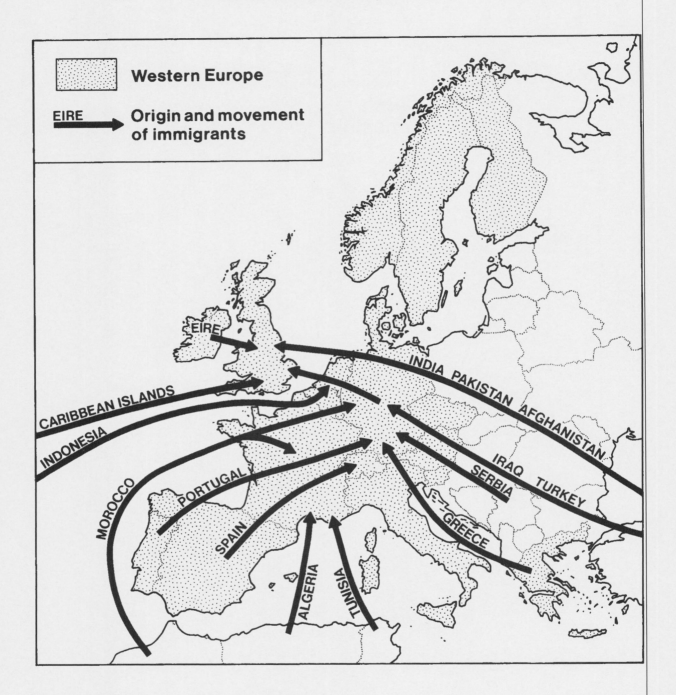

(a) Look at Reference Diagram Q7A above.

Describe the **pattern** of migration into and within Europe since 1990.

7. (continued)

Reference Diagram Q7B: Extract from Newspaper Article

> **Families from troubled countries given asylum**
>
> About 1200 Iraqi and Afghan people were given four-year work permits to live and work in the UK. One refugee said, "I can start work tomorrow. I have useful skills which can help me to provide for my family and put something back into this country."
>
> Some local people say that they are very unhappy about the migrants moving to the UK.
>
> "We thought we'd be safe in this country, but my family are still being persecuted," said one young mother from Kosovo; "Not everyone welcomes us."

(b) Look at Reference Diagram Q7B above.

What are the advantages **and** disadvantages **to migrants** of coming to countries in the European Union, such as the United Kingdom?

4

[Turn over

8. **Reference Diagram Q8: Location of the 10 new Members of the European Union**

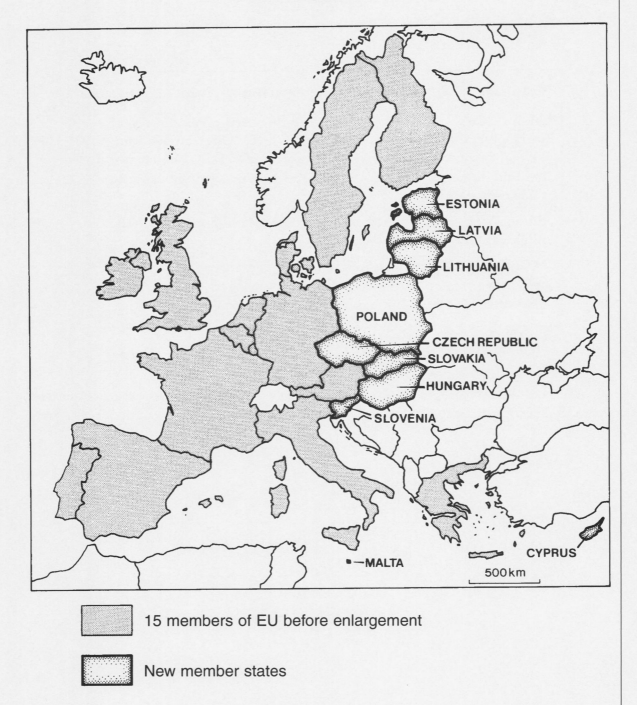

The European Union has been enlarged from its previous membership of 15 countries to a group of 25.

Explain the economic and political advantages to the 10 new countries of joining the European Union.

9. Reference Diagram Q9A: Mount Nyiragongo erupts, 17 January 2002

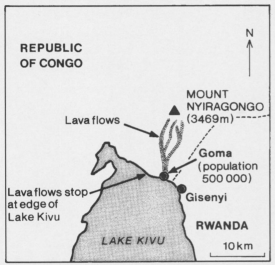

Reference Diagram Q9B

- dozens killed
- crops and farmland destroyed
- lava flows sweep through 14 villages, setting fire to fuel and power stations
- Goma Airport runway blocked by lava
- water supplies cut off
- 10 000 people made homeless
- harbour facilities on Lake Kivu destroyed

Reference Diagram Q9C

Short Term Aid	Long Term Aid
Tents and blankets	Road and bridge repairs
Medicines	New house building
Food supplies	Farming equipment and fertilisers

Look at Reference Diagrams Q9A, Q9B and Q9C.

Following the eruption of Mount Nyiragongo, aid was rushed to the area.

Which type of aid would be best suited to helping the people of this area following the volcanic eruption?

Give reasons for your answer.

5

[END OF QUESTION PAPER]

2005 | General

[BLANK PAGE]

Official SQA Past Papers: General Geography 2005

FOR OFFICIAL USE

KU | ES
Total Marks

1260/403

NATIONAL QUALIFICATIONS 2005

WEDNESDAY, 11 MAY 10.25 AM–11.50 AM

GEOGRAPHY
STANDARD GRADE
General Level

Fill in these boxes and read what is printed below.

Full name of centre

Town

Forename(s)

Surname

Date of birth
Day Month Year

Scottish candidate number

Number of seat

1 Read the whole of each question carefully before you answer it.

2 Write in the spaces provided.

3 Where boxes like this ☐ are provided, put a tick ✓ in the box beside the answer you think is correct.

4 Try all the questions.

5 Do not give up the first time you get stuck: you may be able to answer later questions.

6 Extra paper may be obtained from the invigilator, if required.

7 Before leaving the examination room you must give this book to the invigilator. If you do not, you may lose all the marks for this paper.

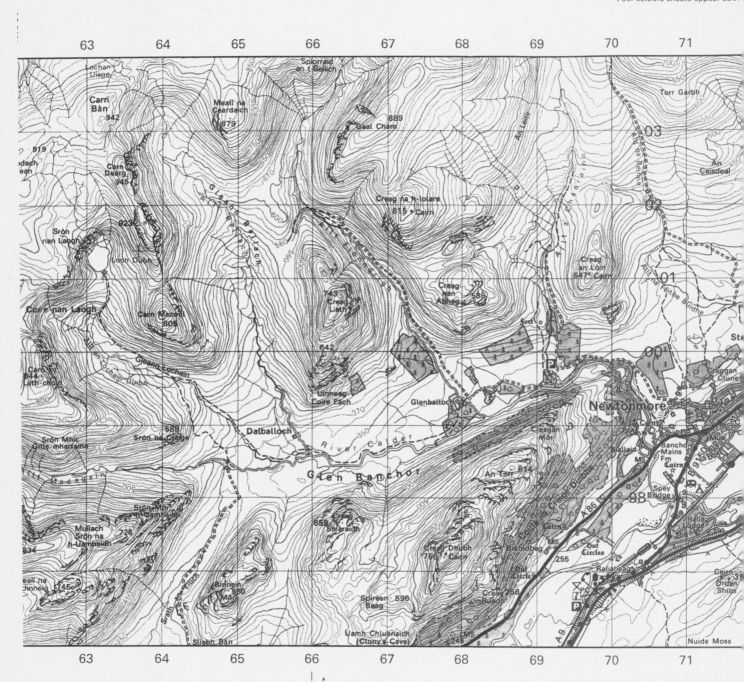

Official SQA Past Papers: General Geography 2005

Extract No 1406/35
1:50 000 Scale
Landranger Series

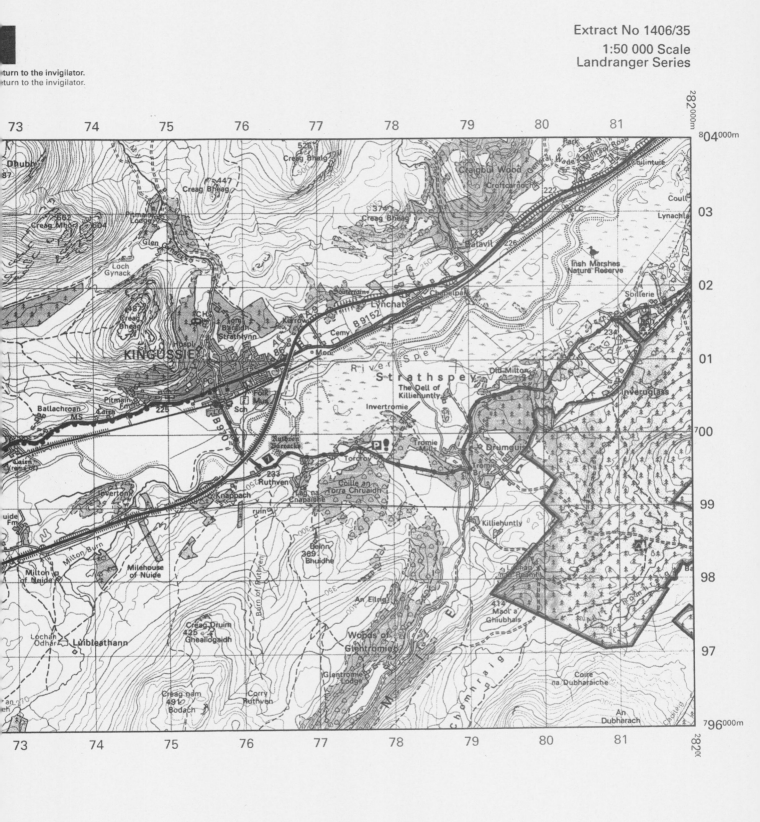

1. Look at the Ordnance Survey Map Extract (No 1406/35) of the Kingussie area.

(a) (i) Match the glacial features in the table with the grid references below. Choose from:

6399   6699   6301   7796.

Glacial Feature	Grid Reference
U-shaped valley with misfit stream	
Corrie with lochan	
Hanging valley	
Truncated spur with crags	

(ii) **Explain** how **one** of the glacial features named in the table above was formed. You may use a diagram(s) to illustrate your answer.

_____
_____
_____
_____
_____

1. (continued)

(b) The River Spey between 720981 and 820039 has a wide flood plain. Study this section of the River Spey. Using map evidence,

(i) give **two** ways in which people have made use of the flood plain;

_____

_____

_____

(ii) give **two** ways in which people have tried to overcome or avoid the problem of flooding on this section of the River Spey.

_____

_____

_____

4

(c) Pitmain Farm is in grid square 7400. What type of farm is this likely to be?

Tick (✓) your choice.

Livestock ☐   Mixed ☐   Arable ☐

Give reasons for your answer.

_____

_____

_____

_____

3

[Turn over

1. (continued)

(d) Match the site descriptions in the table with the settlements shown below:

Insh (8101)   Newtonmore (7199)   Kingussie (7500)   Glenballoch (6799).

Description of Site	Name of Settlement
Between two tributaries to the north of the River Spey	
On land sloping gently down to the northwest, surrounded by forests	
On the floor of a U-shaped valley, next to a tributary of the River Calder	
A tributary of the River Spey runs through the middle of this settlement	

Marks 3

(e) Do you think the area around Newtonmore is suitable for tourism? Using map evidence, give reasons for your answer.

_____

_____

_____

_____

_____

_____

4

1. (continued)

   (f) **Reference Diagram Q1A: Sketch Map showing part of Insh Marshes Nature Reserve**

   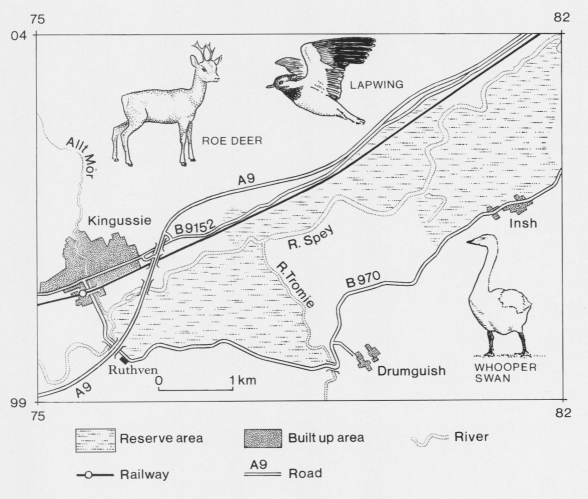

   Look at the Ordnance Survey map and Reference Diagram Q1A.

   The Insh Marshes Nature Reserve extends on both sides of the River Spey from grid square 7699 (Ruthven) to grid square 8103.

   Using map evidence, give advantages and disadvantages of the site of this nature reserve.

   Advantages _____

   _____

   _____

   _____

   Disadvantages _____

   _____

   _____

   4

2. **Reference Diagram Q2: Waterfall**

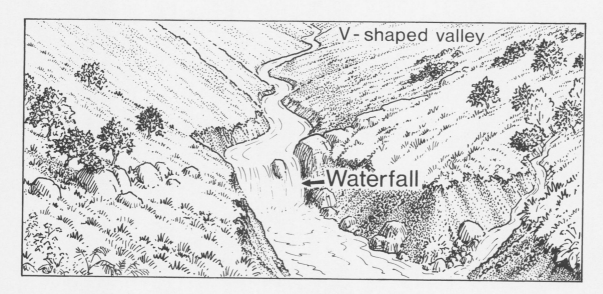

Explain how a waterfall such as the one shown in Reference Diagram Q2 is formed.

You may use a diagram(s) to illustrate your answer.

[Turn over for Question 3 on *Page eight*

3. **Reference Diagram Q3A: Weather Station Sites**

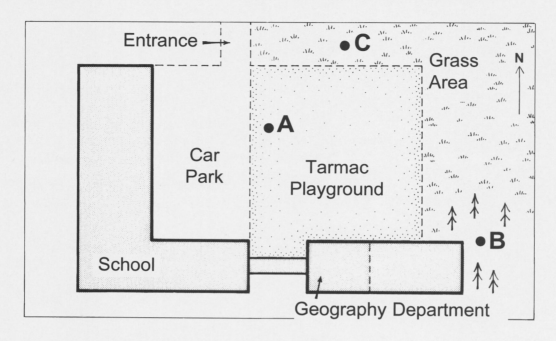

(a) Look at Reference Diagram Q3A.

Do you agree that site A is the best site for the school weather station?

Tick (✓) your choice.

Yes ☐   No ☐

Give reasons for your choice.

_____

_____

_____

_____

_____

3

3. (continued)

Reference Diagram Q3B: Air Masses affecting Britain

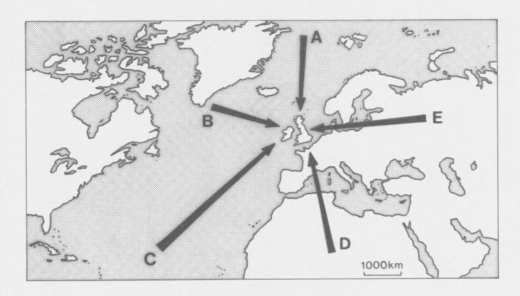

(b) Look at Reference Diagram Q3B.

(i) Describe the type of weather Britain might have with air mass D.

(ii) **Explain** why air mass B would bring cold, wet weather.

[Turn over

4. Reference Diagram Q4A: Selected Climate Regions

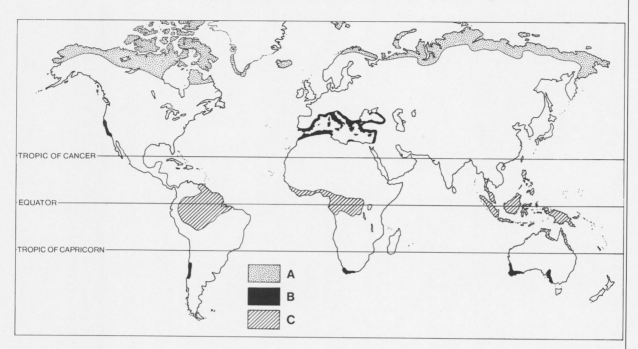

(a) Look at Reference Diagram Q4A above.

Complete the table below by writing in the names of the climate regions.

Area	Name of Climate Region
A	
B	
C	

3

4. (continued)

Reference Diagram Q4B: Desertification in Africa

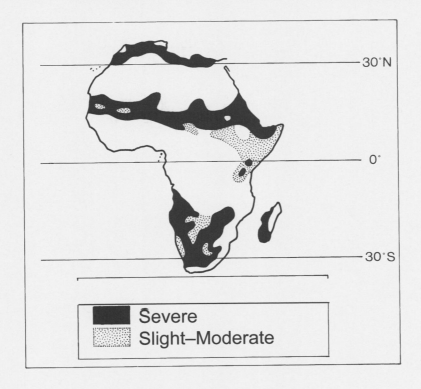

(b) Look at Reference Diagram Q4B.

What are the main causes of desertification?

4

[Turn over

5. Reference Diagram Q5: Land Uses at the Edge of a City

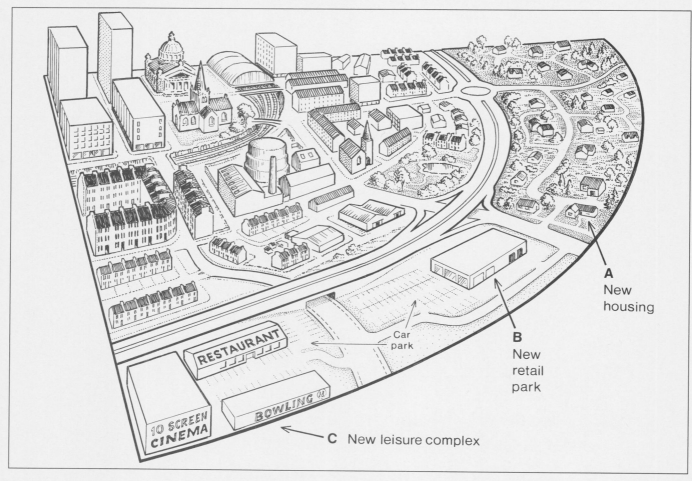

Look at Reference Diagram Q5.

(a) Choose **one** of the main land uses in the diagram—A, B or C.

Chosen land use _____

Explain why the edge of the city is a good location for this land use.

_____
_____
_____
_____
_____

3

5. (continued)

(b) Identify **two** techniques which pupils could use to gather information in a study of an out-of-town shopping centre.

Give reasons for your choice.

Technique 1 _____

Technique 2 _____

Reasons for choice _____

_____

_____

_____

_____

4

[Turn over

6. Reference Diagram Q6: Recent Trends in Farming

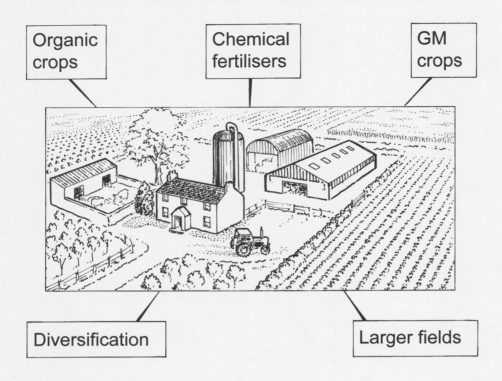

"Recent trends in farming are of great benefit to the British people."
Do you agree? Explain your answer.

4

7. **Reference Diagram Q7: Old Industrial Area**

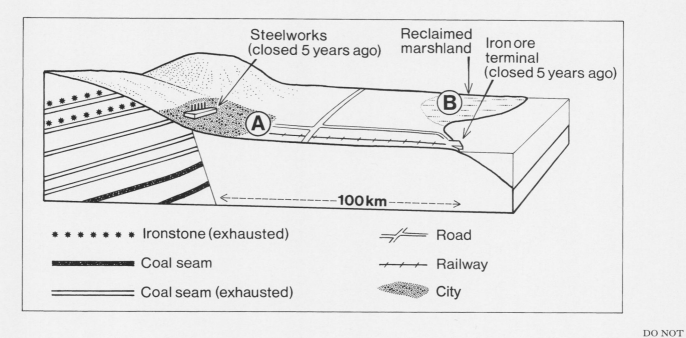

Look at Reference Diagram Q7.

A steel company is considering building a new integrated steelworks at either A or B.

Which site do you think they should choose?

Give reasons for your answer.

Chosen site _____

8. **Reference Diagram Q8A: World Refugees (millions) 1971–2001**

Year	1971	1981	1991	2001
No of refugees (millions)	3	8	18	24

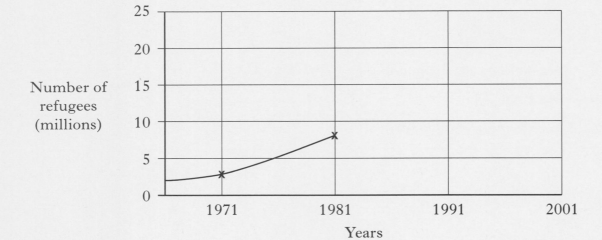

(a) Complete the line graph using the figures from Reference Diagram Q8A.   2

8. (continued)

Reference Table Q8B: Sources of Refugees in Glasgow 2003

Country	Percentage
Afghanistan	6
Democratic Republic of Congo	4
Iran	9
Iraq	7
Pakistan	9
Somalia	9
Turkey	14
Others	42

(b) Suggest **two** other techniques which could be used to process the information in Reference Table Q8B.

Give reasons for your answers.

Technique 1 _____

Technique 2 _____

Reasons _____

4

[Turn over

9. Reference Diagram Q9A: Quotation from Kofi Annan

"Farmers in poor countries must be given a fair chance to compete, both in world markets and at home."

Kofi Annan (United Nations General Secretary)

Reference Diagram Q9B: Maize Trade between USA and Mexico

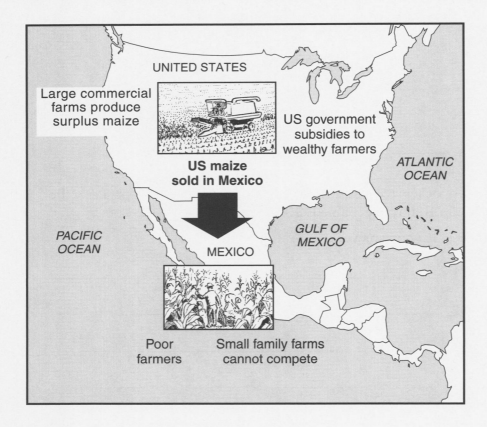

Look at Reference Diagrams Q9A and Q9B.

Explain why Mexican farmers think the maize trade between the USA and their country is unfair.

[Turn over for Question 10 on *Page twenty*

10. Reference Diagram Q10A: Mozambique

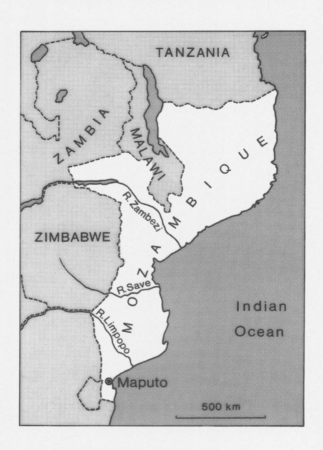

Reference Diagram Q10B: Effects of the Mozambique Floods, February/March 2000

Deaths	1000
Homeless	500 000
Farmland lost	25%
Schools destroyed	600
Estimated cost of rebuilding	$270–$430 million

Reference Diagram Q10C: Types of Aid

Short-term Aid	Long-term Aid
Clean water	Rebuilding homes
Food	Road building
Emergency shelter	Electricity network
Medicines	Building hospitals

10. (continued)

Look at Reference Diagrams Q10A, Q10B and Q10C.

In 2000, two cyclones hit Mozambique causing the Rivers Save, Zambezi and Limpopo to burst their banks. Almost half of Mozambique's land was flooded.

Which type of aid, short-term or long-term, would have been most useful to Mozambique?

Explain your answer in detail.

_____
_____
_____
_____
_____
_____

4

[END OF QUESTION PAPER]

[BLANK PAGE]

2005 | Credit

[BLANK PAGE]

1260/405

NATIONAL QUALIFICATIONS 2005

WEDNESDAY, 11 MAY 1.00 PM – 3.00 PM

GEOGRAPHY STANDARD GRADE Credit Level

All questions should be attempted.

Candidates should read the questions carefully. Answers should be clearly expressed and relevant.

Credit will always be given for appropriate sketch-maps and diagrams.

Write legibly and neatly, and leave a space of about one cm between the lines.

Marks may be deducted for bad spelling and bad punctuation, and for writing that is difficult to read.

All maps and diagrams in this paper have been printed in black only: no other colours have been used.

1. **Reference Diagram Q1**

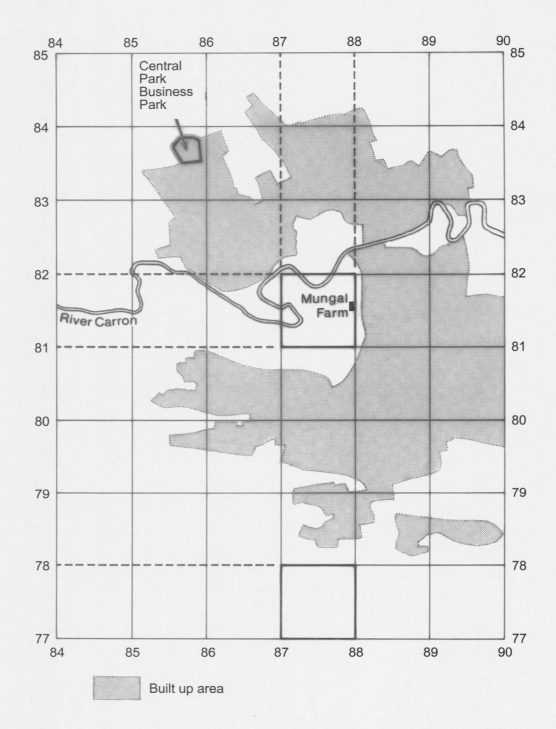

1. (continued)

This question refers to the OS Map Extract (No 1407/EXP349) of the Falkirk Area and the Reference Diagram Q1 on *Page two*.

(a) Describe the **physical** features of the River Carron **and** its valley from Lochlands (859818) to Carron House (897829). [KU: 4]

(b) What is the main present day function of Falkirk?

Choose from:

    industrial   service centre   tourism and recreation.

Use map evidence to support your answer. [ES: 5]

(c) Mungal Farm is found at 880815.

Using map evidence, give the advantages **and** disadvantages of the location of this farm. [ES: 4]

(d) Look at grid squares 8781 and 8777.

Give reasons for the low population density in **each** grid square. [ES: 4]

(e) Use map evidence to **explain** the location of the Central Park Business Park in grid square 8583. [ES: 5]

(f) Do you agree that grid square 8880 contains the CBD of Falkirk?
Explain your answer. [ES: 4]

[Turn over

2. **Reference Diagram Q2: U-shaped Valley**

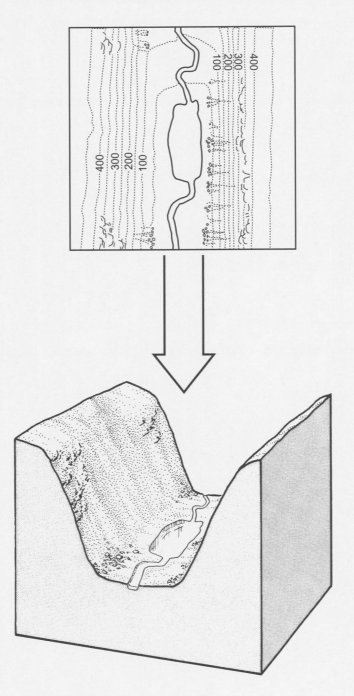

Study Reference Diagram Q2 above.

Explain the formation of a U-shaped valley.

You may use diagrams to illustrate your answer.

3. **Reference Diagram Q3: Land Use Conflict in Loch Lomond and Trossachs National Park**

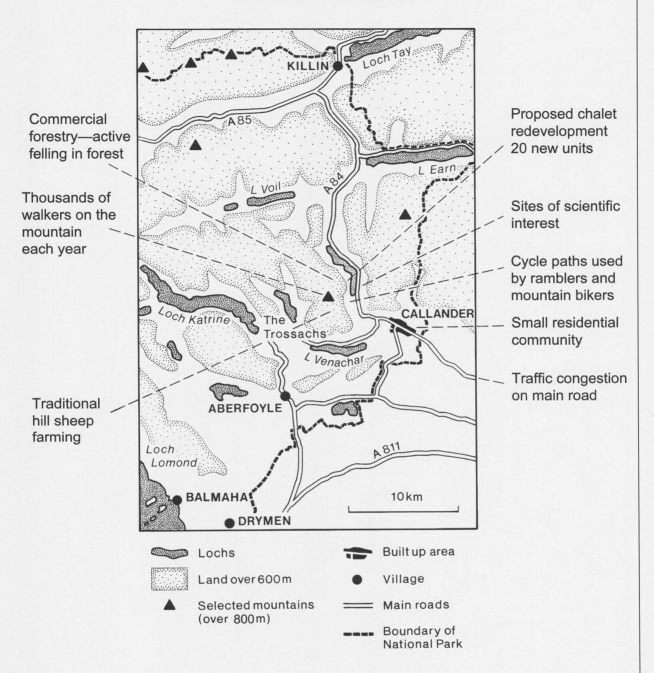

Study Reference Diagram Q3 above.

(a) Select **two** different land uses.

Explain in detail why they may be in conflict with each other.

(b) A group of pupils wants to investigate land use conflicts in the National Park. Describe **two** gathering techniques they could use to collect appropriate data.

Give reasons for your choice.

4. **Reference Diagram Q4A: Weather Conditions in the British Isles**

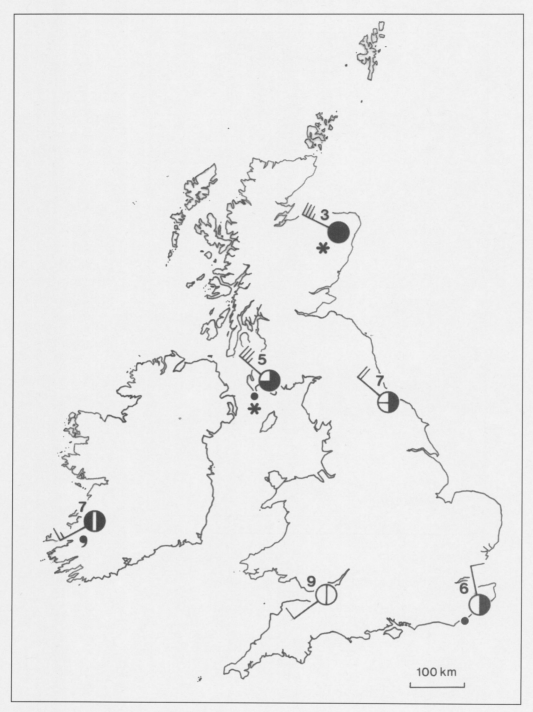

(a) Look at Reference Diagram Q4A.

Compare, in detail, weather conditions in North East Scotland with those in South East England.

4. (continued)

Reference Diagram Q4B: Synoptic Chart for 1200 hrs, 14 November 2004

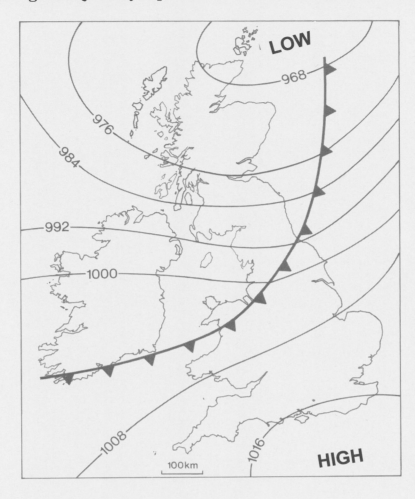

(b) Look at Reference Diagrams Q4A and Q4B.

Do you agree that the weather conditions shown in Reference Diagram Q4A match the synoptic chart in Reference Diagram Q4B?

Explain your answer in detail.

5

[Turn over

5. **Reference Diagram Q5: Llanwern Steelworks, South Wales**

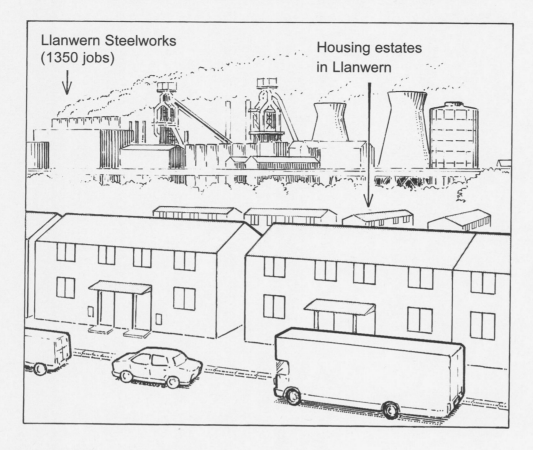

Look at Reference Diagram Q5.

Explain in detail the social, economic and environmental impact of the closure of a large steelworks such as Llanwern on the surrounding area.

6. **Reference Diagram Q6A: Changing Farmscape in UK**

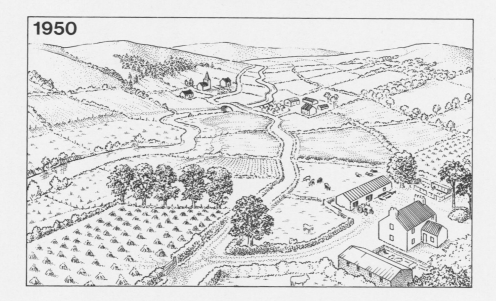

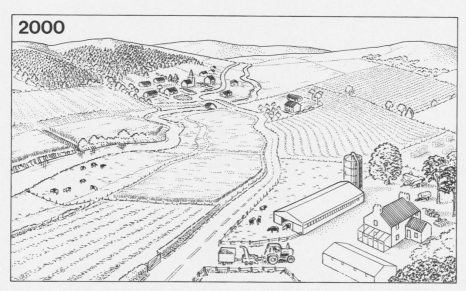

Reference Diagram Q6B: Changes in Farm Size

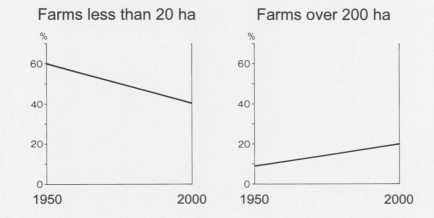

Look at Reference Diagrams Q6A and Q6B.

Explain the advantages **and** disadvantages of the changes shown.

7. **Reference Diagram Q7: Street Maps of two Areas in Greater Glasgow (Scale 1:10 000)**

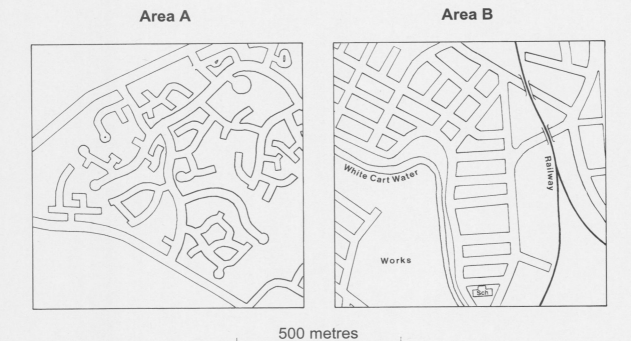

Look at Reference Diagram Q7.

Explain why the urban environments in Areas A and B are different.

Your answer may refer to age, quality of environment, street pattern and location.

6

8. **Reference Diagram Q8: Exports of selected Economically Less Developed Countries (Developing Countries)**

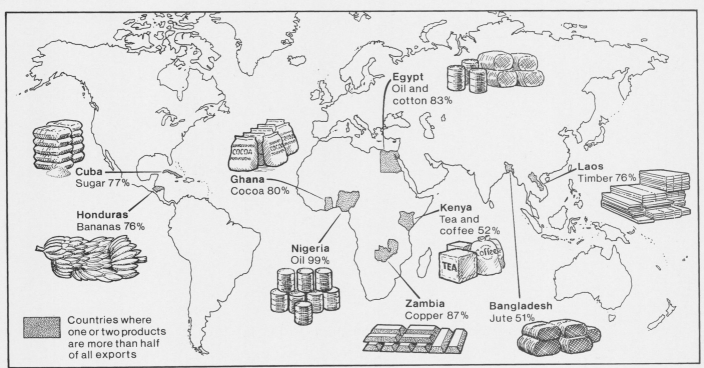

(a) Look at Reference Diagram Q8.

Explain why some Economically Less Developed Countries (Developing Countries) are especially at risk from changing world prices for the goods which they export.

(b) Which other processing techniques could be used to display the export percentage figures shown on Reference Diagram Q8?

Give reasons for your choice of techniques.

[Turn over for Question 9 on *Page twelve*]

9. **Reference Diagram Q9A: Demographic Transition Model**

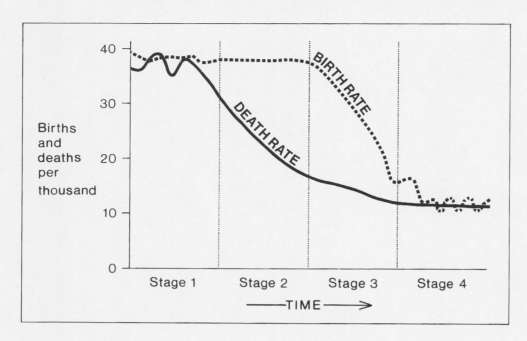

Reference Diagram Q9B: Birth Rate Statistics

Country	Birth Rate/1000
United Kingdom	13
Ethiopia	38

Study Reference Diagrams Q9A and Q9B above.

Many ELDCs* are at Stage 2 in the demographic transition model while EMDCs* are more likely to be at Stage 4.

Give reasons for the differences in birth rates between ELDCs such as Ethiopia and EMDCs such as the UK.

4

* ELDCs = Economically Less Developed Countries
* EMDCs = Economically More Developed Countries

[END OF QUESTION PAPER]

[BLANK PAGE]

Acknowledgements

Leckie & Leckie is grateful to the copyright holders, as credited, for permission to use their material:
This product includes mapping data reproduced by permission of Ordnance Survey on behalf of HMSO © Crown copyright 2005. License no. 100036009.

The following companies/individuals have very generously given permission to reproduce their copyright material free of charge:
Toyota PLC for a photograph (2004 Credit paper p 9).